球兰
资源与栽培

Resources & Cultivation of Hoya Species

黄尔峰　王　晖　贾宏炎　编著
Huang Er-Feng Wang Hui Jia Hong-Yan

深圳市中国科学院仙湖植物园
Fairylake Botanical Garden, Shenzhen & Chinese Academy of Sciences
中国林业科学研究院热带林业实验中心
Experimental Center of Tropical Forestry, Chinese Academy of Forestry
中国球兰网
www.hoyabbs.com

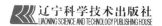
辽宁科学技术出版社
LIAONING SCIENCE AND TECHNOLOGY PUBLISHING HOUSE

《球兰—资源与栽培》编辑委员会
Resources & Cultivation of Hoya SpeciesEditorial Committee

主任：张国宏 蔡道雄
Directors: Zhang Guo-hong, Cai Dao-xiong

副主任：谢良生 张寿洲 邹远军
Vice-Directors: Xie Liang-sheng, Zhang Shou-zhou, Zou Yuan-jun

编委：卢立华 冯昌林 农友 冯海 刘盛楠 陆辉武 周德超 农彬彬 欧振飞 宁永红
Editors: Lu Li-hua, Feng Chang-lin, Nong You, Feng Hai, Liu Sheng-nan, Lu Hui-wu, Zhou De-chao, Nong Bin-bin, Ou Zhen-fei, Ning Yong-hong

编著：黄尔峰 王晖 贾宏炎
Authors: Huang Er-feng, Wang Hui, Jia Hong-yan

序 / Preface

植物引种是一项复杂的工作，需要掌握引种植物的分类、生态、营养和病虫害防治等知识方能为之。本书作者从事球兰属植物引种工作多年，经验丰富，并在这些植物的推广应用上不遗余力。为了能让读者了解球兰属植物的种类、形态和栽培习性，作者特将积累的资料编撰成书。此书是国内第一本介绍球兰属植物的书籍，共收录了栽培较普遍的球兰属植物120种1变形及11个栽培品种。绪言部分，作者参考了近年球兰属植物研究的相关文献，对球兰属植物的分布、分类、生态及形态等内容作了介绍。书中对各种球兰的形态描述文字简练，配以丰富的图片，可以帮助读者认识目前国内栽培的大部分种类。栽培养护部分为球兰属植物的引种和栽培工作提供了参考。书中各种球兰的学名，都经过作者仔细查证并列出了文献引证，故可作为球兰属植物分类学研究人员的参考资料。本书内容翔实，图片精美，图文并茂，实用性强，我乐为之序。球兰属植物种类繁多，形态各异，奇特美观，是优良的观赏植物。希望作者继续进行资料的收集和整理，将更多的种类呈现给读者。

Plant introduction is a profound work, which requires the comprehensive understanding of the plant taxonomy, ecology, nutrition, and pest management. Engaged in the plant introduction of Hoya for many years, the experienced authors are dedicated on the production and application of the plants. For the readers' better understanding of the species, morphology and cultivation of the Hoya species, the author compiled this book based on accumulated information and experience. This is the first book that introduces well-cultivated Hoya species in China, which includes 120 species, 1 forma and 11 cultivars. In the Introduction Chapter, the distribution, classification, ecology and morphology of Hoya is reviewed based on related technical literature in recent year. Combined brief summaries of diagnostic features with stunning color photographs, this landmark book will facilitate the understanding of most Hoya species cultivated in China. The section of Horticulture provides a reference for the introduction and cultivation of Hoya species. All the scientific names of the Hoya species demonstrated in this book are verified by the authors with reference citations, making the book a reliable and useful resource for taxonomists on the genus Hoya. It is my pleasure to write a preface for such a well written, beautifully designed and very readable book including unprecedented photographic coverage. Among the most popular ornamental plants, members of the species-rich genus have apparently rich diversity and remarkable beauty. I hope the authors should keep on pursuing the knowledge of the Hoya species and present more to readers.

李秉滔

华南农业大学 教授
2015年7月于深圳

Li Ping-tao

Professor, South China Agricultural University
July 2015 in Shenzhen City

前言

Foreword

在我国，以兰（Lan）为名的植物很多，它们中的大部分属于兰科 Orchidaceae，但也有一些例外，如木兰 *Magnolia*、马兰 *Aster indicus*、金粟兰 *Chloranthus spicatus* 等，球兰 *Hoya carnosa* 也是其中的一种。东汉的《说文解字》对"兰"的解释是具有香气的草本植物，球兰的花序形似绣球，花芳香，所以这一名称对它的形态和气味作了很好的概括。在1959年出版的《广州植物志》中记录了球兰和凹叶球兰 *Hoya kerrii* 2个种，并提及它们在广州均是常见的观赏植物。该书的主编是著名植物学家、中国科学院院士陈焕镛先生，1984年李秉滔教授发表的椰香球兰 *Hoya chunii* 即以他的姓氏命名。现在球兰一名不仅指 *Hoya carnosa* 这个种，更多的代表了球兰属 *Hoya* 这个类群的植物。球兰属是夹竹桃科 Apocynaceae 中较大的一个属，有300多种。它们的叶片、花序和花的形态变化很大，观赏价值高，是优良的园艺植物。1810年植物学家罗伯特·布朗（Robert Brown）建立球兰属时，便以园艺家托马斯·霍伊（Thomas Hoy）的姓氏命名，似乎预见了球兰属植物将会成为园艺界的宠儿。

本书的第一作者自1995年开始从事球兰属植物的引种与栽培工作。初期由于缺乏参考资料，在种类鉴定和栽培养护上都经过了一番摸索。后来随着经验的积累和设施的改善，收集的种类逐步增加。同时，在国内可供栽培的球兰种类也越来越丰富，估算约有300种及栽培品种的球兰属植物栽培在各地的植物园、苗圃和爱好者家中。专门介绍球兰属植物引种和栽培的网站—"中国球兰网"（www.hoyabbs.com）也为球兰爱好者提供了一个交流的平台。但由于这些球兰植物大多来自国外，在国内依然缺少系统的介绍，这对它们的栽培、养护、推广和应用，以及栽培者的经验交流和分享均有诸多不便。为此，我们根据积累的资料，选出国内常见栽培的球兰属植物120种、1变形和11个栽培品种编撰成此书，并挑选了1200余张彩色照片展示这些球兰的宏观形态特征，读者可以通过这些了解球兰属植物的多样性。同时，本书对这些球兰的栽培习性作了介绍，以供栽培者参考。

华南农业大学的李秉滔教授在此书的编写过程中给予了全面的指导，李教授是夹竹桃科的分类专家，对球兰属植物的形态、分布

In China, many of the Chinese plant's name possess 'Lan'（兰）, in which, most of them are the member of orchid family (Orchidaceae) and few of them are not having, such as 'Mu Lan' (*Magnolia*), 'Ma Lan' (*Aster indicus*), 'Jin Su Lan' (*Chloranthus spicatus*) and so on. 'Qiu Lan' (*Hoya carnosa*) is also a plant name along with 'Lan' in its Chinese name, which belongs to the dogbane family (Apocynaceae). The explanation of Chinese character "Lan"（兰）is present in the book '*Origin of Chinese Characters*' which was written in the Eastern Han Dynasty (121 AD). In this book, Chinese's character was explained this word（兰）as a kind of herbaceous plant with fragrance. In this case, the name 'Qiu Lan' is also summarized well about morphology and smell: ball-shaped (in Chinese 'Qiu' means ball or ball shaped) inflorescence and fragrant flowers. There were two *Hoya* species recorded in '*Flora of Guangzhou*' (published in 1959): *Hoya carnosa* and *Hoya kerrii*, both of them were common ornamental plants in Guangzhou City. The chief editor of this book was the famous botanist Chun Woon-Young, an academician of Chinese Academy of Sciences. Later, *Hoya chunii* was described by Professor Li Ping-Tao in 1984. Now, the Chinese name 'Qiu Lan' refers not only to *Hoya carnosa*, but also to the genus of *Hoya*. *Hoya* is one of the prevalent genera of dogbane family (Apocynaceae), which contains more than 300 species. Leaves, inflorescence and flowers are various in shapes and beautiful. In very particular, these characters make them as an excellent ornamental resource in the horticultural world. In 1810, Robert Brown established the *Hoya* genus and named them after his friend, Thomas Hoy, a horticulturist. His contribution might be the botanist's foresight that *Hoya* plants would become the favorite of horticultural world.

The author of this book began to introduce and growing *Hoya* plants since 1995. Due to lack of references and experience at early stage, the cultivating work has been carried out very slowly. With an accumulation of experience and adequate facility, the collection was gradually increased. At present, an enormous *Hoya* species is now reproduced and available for the plant growers in China. Horticulturists are cultivating nearly 300 species of *Hoya* species in botanical gardens, plant nurseries as well the amateurs' home. In the meantime, the website (www.hoyabbs.com) also provides a place for the communication of *Hoya* amateurs. However, the information of *Hoya* plant's resource and cultivation are still very limited in China. Consequently, with accumulated information besides photos, we described 120 *Hoya* species, and most of them are widely-cultivated in China. Our present book also includes more

和分类历史非常熟悉。他不仅为本书作序，还审阅了书中所有的文字内容，在此向李教授表示衷心的感谢。本书的编研得到了深圳市城市管理局科研项目（项目编号：201415）及中国林业科学研究院热带林业实验中心主持的科研项目（项目编号：2005DIB6J144；SFA2130218）的支持。仙湖植物园的邱志敬博士、黄义钧、郎校安和梁文超在种类收集和栽培养护上给予了帮助，Baskaran Xvier Ravi 博士帮助审阅了本书的英文部分，林漫华和郑曼枞帮助整理稿件和名录；Rungson Phomchareon 先生在引种和照片收集工作上给予了很大帮助，作者在此深表谢意。

　　本书仅收集了常见栽培的球兰属植物 120 种，1 变形和 11 个栽培品种，但随着引种工作的深入和爱好者的增加，将来会有更多的球兰属植物应用在园林和园艺方面。我们也在继续补充收集资料，以期向读者展示更多的种类。由于编写仓促，书中难免有错误和不足，恳请读者批评指正。

than 1200 photos in order to provide each plant's morphological features. We hope that the readers can have a better understanding of *Hoya* species diversity from this book; in addition, it provides the cultivation information of each plant.

We are indebted to our honorable Professor Li Ping-Tao, who has given a comprehensive guidance for species collection, identification and reference collection. We thank the Shenzhen Urban Management Bureau (Project Number: 201415) and Experimental Center of Tropical Forestry, Chinese Academy of Forestry (Project Number: 2005DIB6J144 & SFA2130218) for financial and field work assistance. Many people collaborated and contributed with full of dedication in the preparation of this book, Dr. Qiu Zhi-Jing, Mr. Huang Yi-Jun, Mr. Lang Xiao-An and Mr. Liang Wen-Chao was helped for species collection and plant's cultivation. Our heartfelt thanks to postdoctoral fellow, Dr. Baskaran Xavier Ravi helped to review and preparation of this book. Ms. Lin Man-hua and Ms. Zheng Man-Cong were kindly helped in sorting and arrangement of the name list. Authors are cordially thanks to Mr. Rungson Phomchareon for providing gorgeous photographs.

Most species of *Hoya* species in this book are widely-cultivated in China. By introduction of *Hoya* species and their reproduction, as well amateur's interest, the number are gradually increasing in China. We are still collecting information and photos in order to represent more *Hoya* species. Due to limited schedule, there might be mistakes in this book and criticism from readers is also welcome.

编者
2015 年 7 月于深圳

The authors
July 2015 in Shenzhen City

目录

008	绪言 Introduction	082	椭圆叶球兰 Hoya elliptica
		084	恩格勒球兰 Hoya engleriana
018	环冠球兰 Hoya anulata	086	珊瑚红球兰 Hoya erythrina
020	大花球兰 Hoya archboldiana	088	红副球兰 Hoya erythrostemma
022	阿丽亚娜球兰 Hoya ariadna	090	凹副球兰 Hoya excavate
024	南方球兰 Hoya australis	092	芬莱森球兰 Hoya finlaysonii
026	本格特球兰 Hoya benguetensis	094	费氏球兰 Hoya fitchii
028	贝拉球兰 Hoya bella	096	鞭花球兰 Hoya flagellate
030	双色球兰 Hoya bicolor	098	淡黄球兰 Hoya flavida
032	布拉斯球兰 Hoya blashernaezii	100	香水球兰 Hoya fraternal
034	短翅球兰 Hoya brevialata	102	护耳草 Hoya fungi
036	波特球兰 Hoya buotii	104	黄色球兰 Hoya fusca
038	缅甸球兰 Hoya burmanica	106	光叶球兰 Hoya glabra
040	美叶球兰 Hoya callistophylla	108	球芯球兰 Hoya globulifera
042	钟花球兰 Hoya campanulata	110	球序球兰 Hoya globulosa
044	樟叶球兰 Hoya camphorifolia	112	格兰柯球兰 Hoya golamcoana
046	球兰 Hoya carnosa	114	烈味球兰 Hoya graveolens
048	尾状球兰 Hoya caudate	116	格林球兰 Hoya greenii
050	景洪球兰 Hoya chinghungensis	118	荷秋藤 Hoya griffithii
052	绿花球兰 Hoya chlorantha	120	阿尔孔球兰 Hoya halconensis
054	椰香球兰 Hoya chunii	122	休斯科尔球兰 Hoya heuschkeliana
056	纤毛球兰 Hoya ciliate	124	绒叶球兰 Hoya hypoplasia
058	桂叶球兰 Hoya cinnamomifolia	126	覆叶球兰 Hoya imbricate
060	丘生球兰 Hoya collina	128	基心覆叶球兰 Hoya imbricate f. basi-subcordata
062	革叶球兰 Hoya coriacea	130	红花基心覆叶球兰 Hoya imbricate f. basi-subcordata 'Red'
064	冠花球兰 Hoya coronaria	132	帝王球兰 Hoya imperialis
066	卡氏球兰 Hoya cumingiana	134	粉花帝王球兰 Hoya imperialis 'Pink'
068	丹侬球兰 Hoya danumensis	136	白花帝王球兰 Hoya imperialis 'White'
070	达尔文球兰 Hoya darwinii	138	厚冠球兰 Hoya incrassate
072	厚花球兰 Hoya dasyantha	140	厚冠球兰"日蚀" Hoya incrassate 'Eclipse'
074	戴维德球兰 Hoya davidcummingii	142	厚冠球兰"月影" Hoya incrassata 'Moon Shadow'
076	密叶球兰 Hoya densifolia	144	爪哇球兰 Hoya javanica
078	戴克球兰 Hoya deykeae	146	白花爪哇球兰 Hoya javanica 'Alba'
080	异叶球兰 Hoya diversifolia	148	印南球兰 Hoya kanyakumariana

150	卡斯堡球兰 Hoya kastbergii	218	甜香球兰 Hoya odorata
152	冰糖球兰 Hoya kentiana　150	220	卵叶球兰 Hoya ovalifolia
154	凹叶球兰 Hoya kerrii	222	粗蔓球兰 Hoya pachyclada
156	金边凹叶球兰 Hoya kerrii 'Albo-marginata'	224	巴东球兰 Hoya padangensis
158	银斑凹叶球兰 Hoya kerrii 'Spot leaf'	226	琴叶球兰 Hoya pandurata
160	金心凹叶球兰 Hoya kerrii 'Variegata green'	228	寄生球兰 Hoya parasitica
162	柯氏球兰 Hoya kloppenburgii	230	碗花球兰 Hoya patella
164	裂瓣球兰 Hoya lacunose	232	豆瓣球兰 Hoya pallilimba
166	兰氏球兰 Hoya lambii	234	巴兹球兰 Hoya paziae
168	亚贝球兰 Hoya lanceolata	236	彩芯球兰 Hoya picta
170	棉毛球兰 Hoya lasiantha	238	铁草鞋 Hoya pottsii
172	橙花球兰 Hoya lasiogynostegia	240	猴王球兰 Hoya praetorii
174	白玫瑰红球兰 Hoya leucorhoda	242	伪滨海球兰 Hoya pseudolittoralis
176	黎檬球兰 Hoya limoniaca	244	毛萼球兰 Hoya pubicalyx
178	线叶球兰 Hoya linearis	246	截叶球兰 Hoya retusa
180	罗比球兰 Hoya lobbii	248	反卷球兰 Hoya revolute
182	洛黑球兰 Hoya loheri	250	硬叶球兰 Hoya rigida
184	长叶球兰 Hoya longifolia	252	苏格球兰 Hoya scortechinii
186	洛斯球兰 Hoya lucardenasiana	254	柳叶球兰 Hoya shepherdi
188	香花球兰 Hoya lyi	256	锡亚球兰 Hoya siariae
190	麦氏球兰 Hoya macgillivrayi	258	斑印球兰 Hoya sigillatis
192	红花球兰 Hoya megalaster	260	花葶球兰 Hoya spartioides
194	美丽球兰 Hoya meliflua	262	苏卡瓦球兰 Hoya subcalva
196	美林球兰 Hoya merrillii	264	西藏球兰 Hoya thomsonii
198	小花球兰 Hoya micrantha	266	怀德球兰 Hoya tsangii
200	民都洛球兰 Hoya mindorensis	268	黄结球兰 Hoya vitellinoides
202	蜂出巢 Hoya multiflora	270	瓦氏球兰 Hoya wallichii
204	瑙曼球兰 Hoya naumannii	272	瓦林球兰 Hoya walliniana
206	新喀里多尼亚球兰 Hoya neocaledonica	274	威特球兰 Hoya wayetii
208	秋水仙球兰 Hoya nicholsoniae	276	薇蔓球兰 Hoya waymaniae
210	钱币球兰 Hoya nummularioides		
212	倒卵叶球兰 Hoya obovata	278	中文索引 Index
214	小棉球球兰 Hoya obscura		
216	钝叶球兰 Hoya obtusifolia		

绪言 | Introduction

球兰属（*Hoya*）隶属于夹竹桃科（Apocynaceae）萝藦亚科（Asclepiadoideae），有300余种，分布于亚洲至大洋洲及太平洋岛屿的热带、亚热带地区。球兰属植物多为附生的藤本，茎柔软，缠绕或依靠不定根攀援于乔木或岩石上，稀为附生或地生的半灌木；叶通常对生，有时其中一片退化（如覆叶球兰 *Hoya imbricata*）；该属植物的叶片形态多样，通常革质或肉质，少数种类叶片较薄，表面光滑或粗糙，有时被毛，叶片通常为绿色，有时因生长环境使得一些种类的叶片变为红色、橙红色或紫红色，也有一些种类的叶片表面具有银色、银灰色或橙红色的斑点或斑块。球兰属植物的花序是其区别于其它萝藦亚科植物的主要特征之一，多数种类的花序梗在花或果实脱落后并不随花和果一同脱落，并在下一花期可继续开花。对这种特殊的花序目前还没有统一的定义，常被描述成伞形花序（Umbel）、假伞形花序（Pseudumbel）或伞形状的花序（Umbel-like inflorescences），为便于理解，本书根据其外形统一使用伞形花序（Umbel）描述。与其它萝藦亚科的植物一样，球兰属植物花的构造复杂而精巧，是有花植物中花形态最复杂的类群之一。其花冠辐射对称，有不同的形态和颜色，花冠之上是五角星形的副花冠，通常大而明显，副花冠裂片间为导轨，花粉块藏于其中，而副花冠中心则是柱头，常被雄蕊的附属物覆盖。球兰属植物的雌蕊虽然具2个子房，但其中一个通常不发育，所以果为单生的蓇葖果，即一个果梗上仅具一枚果，而不像其它一些萝藦亚科的植物在一个果梗上具一对蓇葖果（果双生）。

在以往的分类系统中，球兰属为萝藦科（Asclepiadaceae）的一个属。但根据近期植物系统分类学的研究，整个萝藦科应作为夹竹桃科下的一个亚科，本书亦采用此分类观点，将球兰属置于夹竹桃科之下。

球兰属植物的分类

球兰属 *Hoya* R. Br. 由英国著名植物学家 Robert Brown 于1810建立，其属名 *Hoya* 是为纪念他的好友，英国园艺家 Thomas Hoy（图1）。自球兰属建立至今，已陆续发表了600余种，仅2014年

Hoya genus belongs within the dogbane family (Apocynaceae) and milkweed subfamily (*Asclepiadoideae*), which consists of more than 300 species that native to the tropical and subtropical regions of Asia, Oceania and Pacific islands. Most of them are epiphytic herbaceous climbers and their soft stem twine around trees, rocks or climb on a substrate with adventitious roots. Only few of them are epiphytic or terrestrial subshrubs. Almost all species produce two opposite leaves at each node and sometimes with one of the two opposite leaves degenerating (eg. *Hoya imbricata*). In general, leaf shape of different species is extremely variable but usually leathery or fleshy, and very few are thin. The roughness, degree of hairiness and color of leaf are also greatly varied in different species. Moreover, some species have silver, silver gray or orange spots or patches over the leaf surface. The inflorescence of *Hoya* is one of its main features that make it different from other *Asclepiadoideae* plants. Peduncle of most species does not fall off with flowers or fruits, and instead, that can continue to bloom in the next florescence. A special type of inflorescence present in *Hoya* genus, which is often described as umbel, pseudumbel, or umbel-like inflorescences. In this book, umbel is used to describe the inflorescences of *Hoya* according to its appearance. Like other Asclepiadoideae plants, the flowers of *Hoya* are mostly elaborate, complicated and delicate among flowers of angiosperms. In addition, their actinomorphic corolla has various shapes and colors. Staminal corona consists of five erect lobes, usually large and distinct. There are guide rails between corona lobes, in which pollinia hides. Stigma situates in the center of corona and usually covered by anther appendage. *Hoya* species have two ovaries, however, one of which usually grows out into the follicle. Therefore, the fruits are solitary follicle - one fruit on each pedicel, unlike some other *Asclepiadoideae* plants, which have a pair of follicle (binate fruits) on each pedicel.

In previous classification system, *Hoya* genus belongs to milkweed family (Asclepiadaceae). According to recent research on plant systematic, Asclepiadaceae is recognized as a subfamily of dogbane family (Apocynaceae). In this book, we have adopted the same opinion that genus *Hoya* is belonged to Apocynaceae.

Classification

Hoya genuswas named in 1810 by the prominent British botanist, Robert Brown to honor to his friend, the British horticulturist Thomas Hoy (Figure 1). Presently, more than 600 species described; of

图 1. 1810 年 Robert Brown 在新荷兰植物长篇志（Prodromus Florae Novae Hollandiae）中建立了球兰属 Hoya（文献图像来源于 http://www.botanicus.org）

Figure 1. In 1810, Robert Brown established genus *Hoya* in 'Prodromus Florae Novae Hollandiae'. (Image was from http://www.botanicus.org)

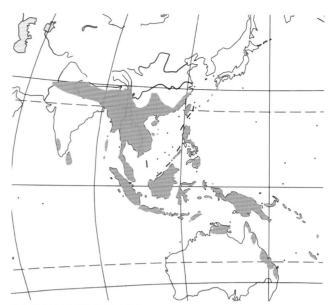

图 2. 球兰属植物的地理分布（根据 Molecular Phylogenetics and Evolution 39: 723, fig.1. 2006. 仿绘）

Figure 2. Geographical distribution of *Hoya* (redraw from *Molecular Phylogenetics and Evolution*, 39: 723, fig.1. 2006.)

就发表了 30 余个新种，这其中虽然存在了大量相同物种被重复发表的问题，但也说明了还有很多种类未被发现。1885 年，英国植物学家 Joseph Dalton Hooker 在他编写的英属印度植物志（The Flora of British India）中，对收录的球兰属植物进行了分类，这是球兰属植物的第一个较完整的分类系统（Hooker, 1883）。此后，许多植物学家均尝试对球兰属植物进行分类修订。但直至现在，该属的分类问题仍然困扰着植物学家，加上每年不断有新种的发表，所以目前球兰属还没有一个完备的分类系统。而对于全世界有多少种球兰，也没有确切的数字，现在被普遍接受的为 300 种左右。

球兰属植物的分布与生态

球兰属植物的分布

1810 年 Robert Brown 在新荷兰植物长篇志（Prodromus Florae Novae Hollandiae）中建立了球兰属，新荷兰是澳大利亚的旧称，为球兰属的自然分布区域。该属是典型的热带亚洲至大洋洲分布的属，分布范围为亚洲至大洋洲的热带、亚热带地区。其分布区的北缘位于我国西藏的南部及东南部，向南可到达澳大利亚的东北部，西缘位于印度南部的西高止山脉附近，向东则可到达南半球的斐济群岛（图 2）。球兰属植物的分布与亚洲及大洋洲的热带雨林和热带季雨林的分布有很高的一致性，大部分的球兰属植物生长在热带雨林和热带季雨林的林缘地带，如森林中的河边和路边，甚至在海岸边，它们通常附生在乔木或灌木的茎干或枝条上，但也有少数较耐寒的种类生长在亚热带的常绿阔叶林中，如我国的贵州、云南等地的常绿阔叶林下。位于赤道附近的苏门答腊岛、婆罗洲岛、苏拉威西岛和新几内亚岛东部是球兰属植物的分布中心，这片区域中球兰的种类最丰富。在垂直分布上，球兰属植物多生长在海拔 1000m 以下的区域，但也有部分的种类可以生长在海拔近 2000m 的地方，如婆罗洲岛和新几内亚岛的高山上均有球兰的分布。

球兰属植物的生态

大部分球兰属植物是热带雨林中的附生植物，通常生长在雨林的林缘地带。它们利用茎和不定根攀附在乔木或灌木的树干上，也有部

them, over 30 new species were recently described in 2014. Even though some of them might be repetitively described, yet there are many undiscovered species. In 1885, British botanist Joseph Dalton Hooker proposed a classification of genus *Hoya* on 'The Flora of British India' which is the first more complete classification system of the genus (Hooker, 1883). Later on, many botanists had tried for the comprehensive taxonomic revision of *Hoya*, but so far it hasn't yet been attempted. Moreover, so many new species are continually described every year. Therefore, both classification systems of the genus and delimitation of species within the genus are still unsatisfactory.

Distribution and Ecology

Distribution

In 1810, Robert Brown described *Hoya* in 'Prodromus Florae Novae Hollandiae'. Novae Hollandiae is the former name of Australia, which is the natural distribution area of *Hoya* species. Genus *Hoya* is typically distributing in tropical and subtropical regions of Asia and Oceania continents. This distribution area ranges north to the west and southeast of Tibetan highlands, south to the northeast of Australia, west to the Western Ghats of southern India, and east to Fiji Islands (Figure 2) which corresponds rather accurately with the distribution of rainforests and monsoon forests in Asia and Oceania continents. Most *Hoya* species grow on the forest edges intropical area, such as river banks, roadsides within forest and even shores, which usually a climb on the stems or branches of trees and shrubs. A few cold-resistant species can be found in evergreen broad-leaf forests in subtropical regions, such as Guizhou and Yunnan of China, etc. The core area of *Hoya* species distributed in Sumatra Island, Borneo Island, Sulawesi Island and eastern New Guinea Island, in addition, a variety of species can be found in these near equatorial regions. About the vertical distribution, most *Hoya* species grow below 1000 m, and only a small number of species can be found at a higher altitude about 2000 m, like mountains of Borneo Island or New Guinea Island.

Ecology

Most *Hoya* species are epiphytic plants living in tropical rainforests, typically on the edges of forests. They climb on the branches of trees and shrubs with their soft stems and adventitious roots, while some species also grow on, usually, calcareous rocks. To adapt this habitat, the morphology, physiological functions and reproductive

图 3. 凹叶球兰 *Hoya kerrii* 的肉质叶片

Figure 3. Fleshy leaves of *Hoya kerrii*

图 4. 爪哇球兰 *Hoya javanica* 的叶片

Figure 4. Thin leaves of *Hoya javanica*

分种类附生在岩石上（主要为石灰岩）。无论在形态结构、生理功能还是繁殖方式上，球兰属植物都具有热带附生植物的特点，从而与栖息地的生态环境相适应。

球兰属植物多为攀援或悬垂藤本。除少数的地生种类外，植株一般具有发达的气生根（不定根），这些气生根除了具有吸收功能外，还有很强的附着能力。植株一般依靠这些气生根攀附在乔木的主干上。发达的气生根还为球兰属植物的无性繁殖提供了便利，当较长枝条折断后，断枝上的气生根可以维持水分供给，从而生长成独立的植株，这一特点让多数球兰属植物的扦插繁殖均有较高的成活率。但也因气生的原因，根需要很好的透气性，如栽培基质紧密或积水导致透气性下降则根部很容易腐烂。

球兰属植物的茎通常较柔软，木质化程度不高。根据茎的攀附能力，大部分的球兰属植物可分为攀援型和悬垂型两类。攀援型的球兰的着生位置通常较低，如乔木的茎干基部或中部，并通过茎和气生根在乔木的茎干上向上攀援。悬垂型的球兰着生的位置通常较高，如乔木的侧枝上。它们的茎通常不攀援，而是向下悬垂在空中。两种类型的球兰均充分利用了雨林中林冠层以下，灌木层以上相对空旷的空间作为栖息地。少部分球兰属植物的茎木质化程度较高，它们的茎既不攀援也不悬垂，而是直立或斜展，呈半灌木状（与茎完全木质化的灌木相区别），这些种类的球兰大部分也附生于林下的树干上或石上，仅有很少的种类为地生植物，如蜂出巢 *Hoya multiflora*。

球兰属植物的叶通常呈肉质（图 3），这也是园艺学家常将其归入多肉植物的原因。这种肉质的叶片与多数多肉植物一样，可以通过景天酸代谢途径（Crassulacean acid metabolism，CAM）在夜间吸收二氧化碳，并储存在叶肉细胞中，肥厚的叶片是二氧化碳的容器。具有这种二氧化碳固定机制的植物通常生长在高温、干旱的环境下，这与球兰属植物生长的雨林环境有很大的差别。但从植株生长的小环境来看，CAM 途径对球兰属植物的生长具有一定的生态学意义。附生的球兰仅能从其附着的树干或岩石上获取水分，这类基质的保水能力差，干湿变化大，一旦干燥时间过长植株就将面临缺水的胁迫，影

pattern of *Hoya* plants entirely feature tropical epiphytic plants.

Most *Hoya* plants are climber or hanging vines except some terrestrial species, and generally, they have well-developed adventitious roots. This kind of root is well in absorption and adhesion, by which the plants can also climb the trunk of trees. Developed adventitious roots also offer great convenience for *Hoya* plant's asexual reproduction. When a long branch breaks off, the adventitious roots on this cutting will continuously provide water supply to grow into a new plant. The propagation of most *Hoya* plants has a high survival rate. However, the roots are aerial, and they need substrates of high air permeability. If the substrate is compact or water-logging, the roots may rot quickly.

The stems of *Hoya* plants are herbaceous and soft in nature. Mostly, *Hoya* plants can be classified into climber or hanging vines. Climbers usually grow in lower position, such as the base or central part of tree trunks, on which the vines climb upward by stems and adventitious roots. Hanging vines grows in higher position, such as branches of trees. The stems of hanging vines usually hang on the air. Both two types inhabit in the empty space between forest canopy and shrub layer. Very few *Hoya* species having medium woody stems, which are not climbing or hanging, but erect or spread, subshrub-like (different from shrubs with entirely lignified stems), and mostly grows on rocks or trunks. Only very limited number of *Hoya* species are terrestrial plants, such as *Hoya multiflora*.

Hoya species usually have succulent leaves (Figure 3). Most of them are considered as succulent plants by horticulturalists. In general, most succulent plant's fleshy leaf absorbs CO_2 through Crassulacean acid metabolism (CAM) at night and stores them in the mesophyll cells. The plants are able to store CO_2 often grow in high-temperature and droughty environment, which is totally different from rainforest environment. However, CAM pathway is necessary for *Hoya* plants because they can only absorb water from the trunks or rocks surface. Due to low water retention capacity, moisture of substrate changes greatly. Once the substrate stays in dry condition for a prolonged period, the growth of plants might suffer from drought. In such case, *Hoya* plants can absorb CO_2 through CAM at night during the evaporation capacity is rather low. Meanwhile, the plants use the stored CO_2 for photosynthesis in a daytime while the evaporation capacity is high. For terrestrial *Hoya* species, such as *Hoya multiflora* and *Hoya Javanica*, due to the

图5. 球兰 *Hoya carnosa* 白色的花冠在夜晚很显眼

Figure 5. White corollas of *Hoya carnosaare* conspicuous at night

图6. 威特球兰 *Hoya wayetiii* 的种子

Figure 6. Plumed seeds of *Hoya wayetii*

响正常生长。在此胁迫下，球兰属植物可以利用 CAM 途径在蒸发量较小的夜间吸收二氧化碳，而在蒸发量大的日间则关闭气孔，同时利用储存的二氧化碳进行正常的光合作用，维持植株的生长。而地生的球兰，如蜂出巢 *Hoya multiflora* 和爪哇球兰 *Hoya javanica*，由于水分供给相对稳定，并不具有肉质叶这一特征（图4）。

所有球兰属植物均为虫媒植物（依靠昆虫传粉），并且目前已知的种类中除了南方球兰 *Hoya australis* 由昼行性的一种蝴蝶协助传粉外，其余的种类均由夜行性昆虫协助传粉，主要是蛾类和甲虫类，可能还包含了部分竹节虫类。常与球兰属植物共生的蚂蚁可能也是其传粉者之一，但所起的作用不大。球兰属植物的花具有吸引夜行性昆虫的特点，它们的花通常在傍晚开放，多数种类的花冠为白色、粉红色、黄绿色或淡黄色，这些颜色在夜晚均较显眼（图5），并且花在夜间产生蜜露和香气。根据观测（Altenburger&Matile，1988），球兰 *Hoya carnosa* 的花释放香气具有很强的节律性，夜间香气浓而白天较淡，中间间隔为 12 小时。球兰属植物花的结构精巧，具有高效的虫媒传粉机制。它们的花粉并不像大多数种子植物的花粉呈分散状，而是聚集成块，称为花粉块（Pollinium），2 枚花粉块与其下的花粉块柄（Caudicle）及着粉腺（Retinaculum）共同组成联合体，称为花粉器（Pollinarium）。每枚雄蕊产生一枚花粉器，并藏于副花冠间的导轨（Guide rail）中。当传粉者停留在花上时，足部容易踩入导轨并带出具黏性的花粉器，从而帮助传粉，这一传粉机制与兰科植物的非常相似。

球兰属植物种子上的种毛可以让种子借助风力远距离传播（图6），甚至可以跨岛屿传播。据文献记载（Bush et al., 1995），1883 年印度尼西亚喀拉喀托岛火山喷发后该岛与附近岛屿上的植物全部死亡，但在 68 年后岛上至少有 2 种球兰分布，这些球兰均为临近的苏门答腊岛和爪哇岛上的种类，它们的种子借助风力传播至喀拉喀托群岛，并在岛上生长。除了依靠风力传播之外，一些种类的种子也会由蚂蚁帮助传播，这些种子被风吹散后，很快会被一些在树上筑巢的蚂蚁收集起来，并种植在蚁巢附近或蚁巢上的"蚂蚁花园"中。蚂蚁收集球兰

relatively stable water supply, the leaves are usually thin (Figure 4).

All *Hoya* species are an entomophilous plant (pollinated by insects). *Hoya australis*, which is diurnally pollinated by a day flying butterfly, while the pollinator of all *Hoya* species could be nocturnal insects, moths, beetles, mainly and probably walkingsticks included too. Ants are live in associate with *Hoya* species that might be a poor pollinator. The flowers of most *Hoya* species open at dusk which combined various flower colors such as white, pink, yellowish green and light yellow (Figure 5). These conspicuous flowers usually produce with a high amount of nectar and strong fragrance, which main play as attractants for the nocturnal insects. By experimental(A Itenburger&Matile, 1988), flowers of *Hoya carnosa* emitted fragrance in a circadian rhythm which is strong in dark period and light in daytime, in other words, flowers timed their emission of fragrance to occur 12 hours after a last emission period. Flowers of *Hoya* species are delicate with efficient pollination mechanism. Unlike most seed plant's disparate pollen, *Hoya* species have cloddy pollen, which also known as pollinium. Each anther produces two pollinia that attached through a caudicle to the retinaculum to form a pollinarium. Pollinatorsbring out the sticky pollinarium while visiting the flower in order to proceed pollination. Moreover, the mechanisms of such a case are similar to many Orchidaceae species.

The plumed seeds of *Hoya* species can be dispersed by wind for long distance (Figure 6), even to an inter-island. The long-distance dispersal was illuminated on Krakatoa Island, where the devastating volcanic eruption in 1883. Due to massive eruption, vegetation of this island and adjacent islands was extinct. Nearly 68 years later, at least two *Hoya* species were native to Sumatra and Java Islands, which recolonized the islands. Probably, seeds were carried out by wind to Krakatoa islands and established new plants there. Besides the mode of wind dispersal, seeds of some *Hoya* species are also dispersed by ants. Once the seeds have been dispersed by wind, thereafter often collected by ants and placed in their nests on trees. The seeds of *Hoya* species germinate and grow up with an accompanied other epiphyte plants such as *Dischidia sp.*, *Dendrobium sp.* and *Pyrrosia sp.* on arboreal ant nests and form a 'ant garden'. The purpose of this seeds collection behavior is various, and generally not for food storage purposes. Mostly, Asclepiadoideae species seeds are dispersed by wind only. Two different modes of seeds dispersals are rarely adopted by some *Hoya* species, which is one of their special ecological features.

图7. 基心覆叶球兰 Hoya imbricate f.basi-subcordata

Figure 7. Hoya imbricate f. basi-subcordata

图8. 达尔文球兰 Hoya darwinii

Figure 8. Hoya darwinii

属植物种子的原因多样，但通常不是为了储藏食物，它们之间的共生关系较复杂。种子具两种传播方式在萝藦亚科的植物中是比较少见的，这也是球兰属植物特殊的生态特点之一，而大多数该亚科植物的种子仅能依靠风力传播。

一些球兰的生长与蚂蚁的关系密切，它们之间存在互惠共生，双方均可从这种关系中获利。除了生长在蚁巢上的"蚂蚁花园"之外，一些球兰还会通过特化叶片，并在叶片与其附着的树皮间（如基心覆叶球兰 Hoya imbricate f.basi-subcordata，图7），或者叶片与叶片间（如达尔文球兰 Hoya darwinii，图8）形成小室吸引蚂蚁前来筑巢，它们的根则在蚁巢中生长。在这种共生关系中，蚂蚁筑巢和捕食可以为球兰带来额外的养分，并帮助球兰传播种子及驱赶采食的昆虫。而球兰除了为蚂蚁提供巢室外，还分泌蜜露作为蚂蚁的食物。同时，球兰的根也可以帮助稳固蚁巢的结构。球兰与蚂蚁之间的共生关系较复杂，据研究显示（Wanntorp et al, 2006），为了适应蚁栖这一生态习性，与蚂蚁共生的球兰在形态上至少经历了三次进化。

球兰属植物的栽培

了解球兰属植物的生态特点对栽培该属植物有很大帮助。大多数球兰生长在低海拔雨林的林缘或林下，附生在乔木或灌木的树干上。这里的气候温暖潮湿，由于林冠层的遮挡，光照也较温和。而有些种类虽然分布在热带，但却生长在海拔较高的地区，通常在山地的云雾带中，这里空气湿度较高但气温较低，并且昼夜温差很大。由于不同种类不同的生态适应性，所以栽培球兰时，需要尽量模仿该种植物的生长环境，才能使植物生长得更好。

光照：很少球兰能耐受阳光的直接照射，强烈的光照会灼伤它们的叶片，但如果栽培地的光照太弱又不能支持它们的生长。如果在室内栽培，理想的地点是南向的窗边或阳台内侧。如果露天栽培，则需要适当的遮阴以避免阳光直射。一些球兰的叶片在较强光照下会变为红色或橙红色，显得较美观。这主要是由于植物叶片中产生了较多花青素和胡萝卜素等色素，以抵抗光照对叶片带来的伤害。

温度：多数球兰分布在热带地区，这些地区全年高温，温度变化

Many Hoya species are growing in association with ants. Some species have specialized leaves to form a cavity between leaves and barks of stem (Hoya imbricate f. basi-subcordata, Figure 7) or between leaves (Hoya darwinii, Figure 8). These cavitie sadapt to housing ants, in which the plant will form roots. On this type of mutualism, the plant can benefits from the extra nutrition which obtained from the organic material collected by ants, and drive foraging insects as well, while the plant provides shelter and food in the form of nectar for ants. Moreover, plant roots can also stabilize the nest structure. The symbiotic relationship between Hoya species and ants is complicated. Ant-symbiosis have experienced at least three times of evolution within Hoya.

Cultivation of Hoya plants

By knowing the ecological character of Hoya species is helpful to cultivate them. Most Hoya species survive under edges or canopy of lowland rainforests and climbing on the trunks of trees or shrubs, which is warm and humid as well. The sunlight is much tenderer along with the shading of forest canopy. However, some Hoya species are distributed in tropical regions with high elevation. Generally, these regions are located on the fog belt on mountains along with high humidity, relatively low temperature than lowlands and large diurnal amplitude. Due to various ecological adaptations of different species, it is necessary to imitate the growing environment of certain species for their better growth.

Lighting: Only a few Hoya species grow well in direct sunlight, and most species tend to get burnt, while fewer amounts of sunlight not supporting the plant normal growth. If Hoya plant is grown interior, the suitable place is southward window-side or balcony inside. If the plant is grown outdoors, proper shading is necessary to shield from direct sunlight. Leaf of some Hoya species will turn red or orange due to strong light and make the plant colorful. So that, more amounts of pigment (such as anthocyanin and carotene) are produced in leaf cells in order to resist strong sunlight.

Temperature: Most Hoya species distributed in tropical regions with warm and stable climate, which makes them mostly nonhardy and need to keep warm while the temperature drops to below 10℃. Furthermore, some species will suffer from chilling damages while the temperature drops to 15℃ below. If these species are grown in cooler place, moisture restriction can help them to resist cold injury. Proper irrigation induces the plant in a drought-resistant condition help to

小。所以球兰属植物大多不耐寒冷，多数种类当温度低于10℃时需要保温，有些种类甚至在温度低于15℃时即受寒害。控制水分可以帮助球兰抵抗低温，所以低温来临前可适当减少浇水，让植株进入抗旱的状态，有助于安全越冬。这种方法使用在具肉质叶片的种类上更为明显。有些球兰对每日气温变化有着苛刻的要求，通常需要昼夜温差较大的环境才能健康生长，如绒叶球兰 Hoya hypolasia 仅在昼夜温差达10℃以上（日间25~30℃，夜间15~20℃）的环境下才会开花。目前已知最耐低温的球兰是原产我国的香花球兰 Hoya lyi，其植株甚至可以耐受短时 -5℃ 的低温，它也是地理分布最北的球兰之一。

湿度：多数球兰喜欢湿润的气候，虽然它们也能耐受一定的干旱，但湿润的气候是理想的生长条件。对于干旱的抵抗，具肉质叶片的球兰由于具 CAM 途径，抗旱性明显强于叶片较薄的种类。如果栽培的球兰出现缺水导致叶片萎蔫，不可以过度浇水，这样容易导致根部死亡。有效的方法是先浇少量的水，并将植株用塑料袋包裹好，袋内的高湿度可防止叶片继续失水，少量的水分也不会伤害根系。如此，植株就有机会调整体内水分含量，进入正常的水分代谢状态，从而恢复正常生长。

栽培基质：多数球兰为热带地区的附生植物，它们生长的地区虽然终年多雨，但由于雨水在雨后会及时排干，所以这些球兰的根均不耐水湿。对于栽培基质的选择，也需要以疏松、透气为主。但是如果过分追求透气性，易造成保湿性下降，导致根部缺水。所以必须要求基质在透气和保湿上达到平衡，才利于植株的生长。通常的要求是选择颗粒间空隙较大，利于空气流通和水分排出，但颗粒本身却可保持湿润的基质。50%的泥炭土、20%的碎瓦片与30%的珍珠岩是比较理想的栽培基质，如需增加保水性，则可适当提高泥炭土的比例，反之则提高珍珠岩的比例。颗粒状的椰壳也是理想的栽培基质，甚至不需要配以其他基质即可用于球兰的栽培。但需注意的是目前市面上的一些椰壳颗粒含盐量较高，需要反复浸泡冲洗才可用于栽培。

肥料：适当施肥可促进植株的生长，特别是栽培基质本身缺少肥力的情况下更需要定期施肥保持植物营养供给。球兰属植物对肥料没

resist at a cold temperatures. This method is more effective and adoptable for *Hoya* species with fleshy leaves. Some *Hoya* plants need subtle requirements for temperature fluctuation and usually growing healthily in the huge temperature difference between day and night. For instance, *Hoya hypolasia* only blooms at 10 ℃ above (25-30 ℃ in day time and 15-20 ℃ in night time). Among *Hoya* species, the most cold-resistant species is *Hoya lyi*, which is native to China and even can bear a temperature as low as -5 ℃ in 3-5 days. *Hoya lyi* is also one of the *Hoya* species, which distributed in most northward.

Humidity: Most *Hoya* species like humid environment. Although they are growing at drought condition, humid environment is preferable. With the help of CAM photosynthesis, drought resistance of fleshy leaf *Hoya* species is much better than thin leaf *Hoya* species. In water deficit condition, leaves are wilted, and irrigation should be avoided to protect roots from rot. The best efficient way is by providing less water content and wraps the plant with plastic bag, in which the high humidity can prevent the leaves from water loss. Moreover, the plant gets adequate water and performs a normal state of water metabolism for normal growth.

Substrate: Most *Hoya* species are epiphytes and growing in tropical regions with plenty of annual rainfall. Due to rapid drains out of rainwater from the barks where *Hoya* plants adhere and the roots are usually not resistant to water logging. Therefore, loose and ventilate substrate balanced between air permeability and moisture retention capacity is better for cultivating *Hoya* plants. One of the most suitable substrate recipes is 50% peat soil; 20% broken tiles and 30% expanded perlite. By a proper increase with the proportion of peat soil, plants can improve their water-retention capacity. On the contrary, increase the proportion of expanded perlite. Granular coconut husks are also a kind of good substrate, which can be used without any other substrate. Some granular coconut husks commercially available on the market which contains high salinity, and need repeated soaking before use them.

Fertilizers: Regular fertilization is required to maintain nutrition supply and promote the growth of *Hoya* plants instead of the substrate is infertile. Generally, ordinary compound fertilizer can supply the nutrition demand of most *Hoya* species. For those *Hoya* species with long stems and large climbing areas, regular foliage spray is necessary to healthy growth of plant. Furthermore, controlled-release compound fertilizer is a good choice to increase

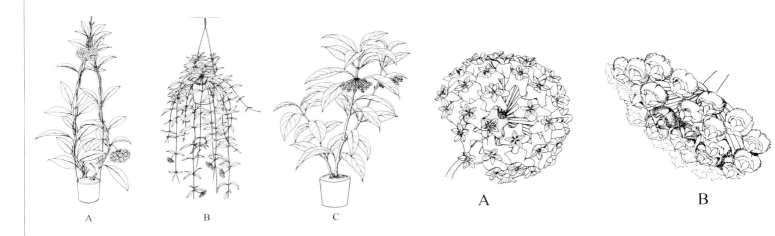

图 9. A. 攀援藤本；B. 悬垂藤本；C. 半灌木（李志民绘）
Figure 9. A. Climbing vine; B. Hanging vine; C. Subshrub (drawn by Li Zhi-Min)

图 10. A. 伞形花序呈球状；B. 伞形花序呈平头状（李志民绘）
Figure 10. A. Umbel convex; B. Flat-topped umbel (drawn by Li Zhi-Min)

有特殊的要求，普通的复合肥即可满足需求。而对于茎很长，攀援面积大的球兰，定期喷洒叶面肥能促进其更好地生长。此外，为了延长施肥周期，具有缓释功能的复合肥也是不错的选择。

繁殖：球兰的传粉需要昆虫协助，可能是缺少传粉者，所以栽培的球兰很少结果。目前球兰的繁殖以无性繁殖为主，方法主要有压条法和扦插法。压条法主要用于茎嫩、叶片较薄、不耐旱的种类。将水苔和珍珠岩混合的基质放置于植株周围并保持基质湿润，将茎略压入基质中，根系就会从覆盖处长出。待根系生长成熟后，便可将此枝条从母株上割离，形成新的植株。扦插法则适用于大部分的球兰种类。对于插条的选择，一般选用成熟的枝条，但木质化的枝条由于较难生根所以一般不选择。插条在扦插之前需要进行适当的修剪，一般叶片较小的种类每个插条留两个节，而叶片较大且肉质的种类留一个节即可。由于节下的节间部分较容易生根，所以在修剪插条时，节下部可留 2~3cm 的节间。插条的新芽会从节上伸出，故插条节上部的节间可以去除。将插条插入基质时需注意方向，如果方向相反插条即死亡。同时，节间需置于基质表层或浅层处，包埋过深插条容易腐烂。有时也可将插条平铺在基质上，上略覆盖基质即可。扦插好的插条需置于湿度较高的空间，以防止插条失水。也可用塑料袋将插条及容器包裹起来保湿，待根充分发育后再除去塑料袋。一些生根能力强的球兰，如凹叶球兰和球兰等，将插条置于水中也会生根，但后期移苗时由于不带基质，需注意保护幼嫩的根部。

the fertilization period.

Reproduction: All *Hoya* species are entomophilous plant. The pollination process could not be finished without the help of pollinator. Layering and cutting is the main reproductive method to cultivate *Hoya* plants. Layering is applicable to less-drought-resistant *Hoya* species with tender stems and thin leaves. Along with the moisture, put the substrate of sphagnum moss and perlite around the plant. Then, slightly press the stem into the substrate and the plant will form roots. When the root steady grows, this branch can be cut off from mother plant for further establishment of a new plant. Usually, mature branch is prepared for cutting which applicable to most *Hoya* species. The lignified mature branch should be avoided for cultivation because of failure in root formation. Cuttings need proper trimming in advance: For those species with small leaves, two nodes should be remained within each cutting. For those species with large and fleshy leaves, usually one node is sufficient. The roots of cuttings often raised from lower part of internode. 2-3cm lower internode can be left while trimming the stem cutting. Meanwhile, buds will spread from the leaf axil of the nodes, and the upper internode of the cutting can be removed. The internode of the cutting should be placed over the surface layer when insert the cutting into substrate or otherwise, it may probably rot. Another way is spread the cutting on the substrate and cover with a thin layer of substrate. Later, keep the container in a high humidity place to avoid water loss. The cuttings along with container could be wrapped with transparent plastic bag to keep moisture. The plastic bag can be removed after full developed roots of cutting. Some *Hoya* species such as *Hoya kerrii* and *Hoya carnosa* can take roots in the water. While transplanting the cuttings into a pot, the tender roots should be protected carefully.

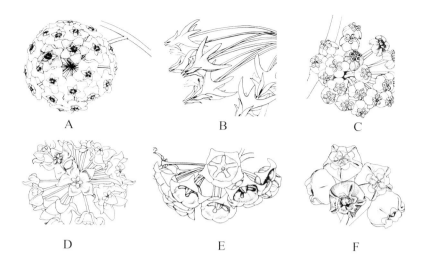

图 11. A. 花冠平展；B. 花冠向背面反折；C. 花冠向背面反卷；D. 花冠内弯呈爪状；E. 花冠钟状；F. 花冠坛状（李志民绘）
Figure 11. A. Corolla spreading; B. Corolla reflexed; C. Corolla revolute; D. Corolla incurved (claw-like); E. Corolla campanulate; F. Corolla urceolate (drawn by Li Zhi-Min)

本书解读

学名

本书所收录种类的学名及引证文献主要来自密苏里植物园植物资料库（Tropicos）及国际植物学名索引（International Plant Name Index, IPNI）。由于本书仅作球兰属植物的种类及栽培介绍，故该名称的异名及普用名均未列出。所收录种类的排序由种加词的首字母排序。

中文名

由于栽培的球兰多数来自国外，此前在国内并没有统一的中文名称，但随着球兰栽培者的增多，多数的种类目前均有普遍接受的中文名称。本书所使用的中文名，尽量以普遍接受的名称为主。当遇到中文名与植物学名之意义有明显差异时，本书则根据学名的来源或植物特点重新拟定中文名，故凡中文名后带有"新拟"的，均为本书作者新拟定之中文名。同时，将原中文名置于新拟名称之后，避免读者误解。本书末附有中文名称索引，方便读者查阅相关种类。

分布

为该植物的自然分布地，信息主要来自相关参考文献及野外观测。由于调查工作还不够深入，所以一些种类的分布可能比书中列出的要广。

形态特征

描述了该植物的植株形态、茎、叶片、花序和花的宏观形态特征，供读者辨识或选择栽培本书所收录的种类。为便于鉴别和描述，本书对植株、花序和花冠的形态作了大致的分类：

植株形态： 分为攀援藤本、悬垂藤本和半灌木3种（图9）。

花序： 分为球状和平头状2种，另未提及形状的花序，均为略呈凸面的花序（图10）。

花冠： 分为花冠平展、花冠向背面反折、花冠向背面反卷、花冠内弯呈爪状、花冠钟状及花冠坛状6种（图11）。

Reading this book

Scientific Name

Scientific names and citation of the *Hoya* species in this book follow rules of botanic nomenclature and mainly all the references obtained from the Missouri Botanical Garden Library (Tropicos) and International Plant Name Index (IPNI). Species are represented in alphabetical order.

Chinese Name

Most *Hoya* species cultivated in China were introduced from abroad and not having common Chinese's name in early period. Due to their popularity horticultural field of China, most *Hoya* species were given common Chinese name by the Chinese growers recent year. The Chinese names of *Hoya* species in this book are mainly gathered from theplant growers.

Distribution

The natural distribution of *Hoya* species was mainly collected from related references and field observation. Because there are many regions sampling effort has been not enough, and some species distribution range may be wider than indicated here.

Morphology

Macro-morphological characteristic which include plant form, stem, leaf, inflorescence and flower of the species was described for readers in order to identify or to select *Hoya* species. For an easy identification, based on three groups of general classifications such as plant form, inflorescence and corollas was given throughout this book.

Plant form: A group of three major plant forms was selected: climbing vine, hanging vine and subshrub (Figure 9).

Inflorescence: There are two types of inflorescence mentioned: convex and flat-topped (Figure 10). In addition, for those species inflorescence shapes are not mentioned, all of them have slightly convex inflorescence.

Corolla: A group of six major types of corolla are referred: spreading, reflexed, revolute, incurved (claw-like), campanulate and urceolate (Figure 11).

栽培习性

介绍了适合植物生长的光照、温度和湿度等条件。由于扦插繁殖是目前球兰属植物主要的繁殖方式,在这里也对该种植物枝条的扦插成活率作了简要的介绍。

备注

记载了该种球兰的名称来源及植物特点等信息,供读者参考。

光照需求

是适宜该种植物生长的光照强度。由于植株对光照具有一定的适应范围,故本书仅将植物的光照需求分为5等,由少到多表示光照强度逐渐增强,图标所对应的光照强度为:

☀ : 20~30% 的全日照;

☀ ☀ : 30~50% 的全日照;

☀ ☀ ☀ : 50~70% 的全日照;

☀ ☀ ☀ ☀ : 70~90% 的全日照;

☀ ☀ ☀ ☀ ☀ : 90~100% 的全日照。

栽培者可以根据每种球兰的光照需求选择植株的栽培地,也可根据该需求选择相应的遮阳网为植株遮阳。值得注意的是,几乎没有球兰能耐受100%的光照,但一些种类需要强烈的光照才能促进开花。故光照需求为90~100%的球兰通常仅在花期之前置于全日照下催花,待植株开花后即可移至遮阳处栽培。

养护难度

根据本书作者总结在福建漳州、广东深圳和广西南宁多年栽培及观察的经验,并综合其所需的栽培条件给出的栽培养护难易度建议。难度共分为5等,由少到多表示栽培养护难度逐渐增加,图标所对应的难度为:

✎ 养护容易,该种球兰的适应性很强,对栽培条件要求很低,粗放养护即可正常生长;

Habits

The information of habits includes light, temperature, humidity and other environmental conditions suitable for plant's growth. The information is based upon the observation of the plants grow in the nursery in southeast, south and southwest of China. A brief introduction of the cutting's growing and their survival rate was also provided for growers to reproduce *Hoya* species.

Note

Other information such as origin of the plant's scientific name or special characteristics also mentioned here for reference.

Lighting requirement

The suitable light intensity was given for plant rapid growth. Because most *Hoya* species have an adaptive range for lighting, and their lighting requirements are divided into five grades. The following icons represent the increment of light intensity:

☀ : 20-30% of full sunlight;

☀ ☀ : 30-50% of full sunlight;

☀ ☀ ☀ : 50-70% of full sunlight;

☀ ☀ ☀ ☀ : 70-90% of full sunlight;

☀ ☀ ☀ ☀ ☀ : 90-100% of full sunlight.

According to the lighting requirement, growers can choose the growing place for each species, or have to choose shade net to adjust light intensity. Noteworthy only a few *Hoya* species grow well in 100% full sunlight, while some species require strong sunlight to promote flowering. Those species on 90-100% of full sunlight growers can place them in full sunlight to induce flower. Then, move them to shadings after forming inflorescences.

Maintenance difficulty

The suggestions on maintenance difficulty of each species are represented here, which based on observation and cultivation experiences in nursery. The authors to this book have three *Hoya* plant nurseries separately in southeast (Zhangzhou City), south (Shenzhen City) and southwest (Nanning City) of China. The successful rates of cultivation conditions to each *Hoya* species and five different grades were presented. The gradually increased number of icons represents the increment of maintenance difficulty, as follows:

🌱: 较易养护，该种球兰的适应性较强，对栽培条件要求较低，仅需满足基本的遮阳及给水即可正常生长；

🌱🌱: 正常养护，该种球兰有一定的适应能力，但需满足其生长的光照、温度、湿度及水分供给等基本条件，否则植株生长不佳；

🌱🌱🌱: 不易养护，该种球兰对栽培环境要求较苛刻，光照、温度、湿度均需符合该种植物的要求才能正常生长，对栽培基质也有较高的要求，并且需要根据基质及环境湿度适量给水，过湿则易烂根，过干则植株萎焉；

🌱🌱🌱🌱: 很难养护：对栽培环境要求很苛刻，稍有缺失即易致植株死亡，有时需要诸如昼夜温差大、基质干湿循环等特殊的栽培要求。这类球兰通常是生境较为特殊，分布区域狭窄的种类。

🌱: Very easy to grow. Plant is well-adapted and has low demand for cultivation conditions. The plant can perform regular growth under extensive cultivation.

🌱🌱: Easy to grow. Plant has good adaptability and less demanding for cultivation conditions. Plants can keep their regular growth by proper shading and irrigation.

🌱🌱🌱: Normal to grow. Plant has adaptability, but the basic conditions such as lighting, temperature, humidity, irrigation supply and other needs are necessary to its growth. Or otherwise, it leads to poor growth.

🌱🌱🌱🌱: Uneasy to grow. Plant needs subtle requirements such as lighting, temperature and humidity to grow normally. Irrigation is based on the substrate recipe and the environmental humidity. Too much irrigation will probably lead to rotten roots, while dryness will lead to wilting.

🌱🌱🌱🌱🌱: Difficult to grow. Plant needs proper cultivation environment. Absence of subtle requirements will lead to plant death. Sometimes, the conditions include high-temperature difference between day and night, wetting and drying cycle, and other special requirements. These species usually grow in special habitat, and their distributions are too narrow.

主要参考文献
References

蒋英, 李秉滔, 1977. 萝藦科: 中国植物志 63, 249-575.
李秉滔, 陈锡沐, 庄雪影, 2011. 萝藦科: 广西植物志 3, 866-946.
Altenburger, R., Matile, P., 1988.Circadian rhythmicity of fragrance emission in flowers of *Hoya carnosa* R. Br. Planta 174(2), pp. 248-252.
Brown, R., 1810. Asclepiadeae. In: Brown, R., Prodromus Florae Novae Hollandiae, 458-464.
Bush, M.B., Whittaker, R.J., Partomihardjo, T., 1995. Colonization and succession on Krakatau: an analysis of the guild of vining plants. Biotropica 27, 355-372.
Hill, K.D., 1988. A revision of *Hoya* (Asclepiadaceae) in Australia.Telopea Journal of Plant Systematics 3(2), 241-255.
Hooker, J.D., 1885. Asclepiadeae. In: Hooker, J.D. (Ed.), Flora of British India 4, pp. 1 - 78.
Kleijn, D., Van Donkelaar, R., 2001. Notes on the taxonomy and ecology of the genus *Hoya* (Asclepiadaceae) in Central Sulawesi.Blumea 46, 457-483.
Kunze, H., 2005. Morphology and evolution of the corolla and corona in the Apocynaceae s. l. Botanische Jahrbücher für Systematik, Pflanzengeschichte und Pflanzengeographie 126(3), 347-383.
Kunze, H., Wanntorp, L., 2008. The gynostegium of *Hoya spartioides* (Apocynaceae-Asclepiadoideae): A striking case of incongruence between molecular and phenotypic evolution. Organisms, Diversity & Evolution 8, 346-357.
Li, P.T., Gilbert, M.G., Stevens, W.D., 1995. Asclepiadaceae. In: Wu Z.Y., Raven P.H. (Ed.), Flora of China 16, 189-270.
Rodda, M., Ercole, E., 2014. *Hoya papaschonii* (Apocynaceae: Asclepiadoideae), a new species from southern Thailand with a peculiar corona. Phytotaxa 175 (2), 97-106.
The Plant Names Project.,2015. International plant names index. Published on the internet: http://www.ipni.org.
Tropicos., 2015. Missouri Botanical Garden.Published on the internet: http://www.tropicos.org.
Wanntorp, L., Kocyan, A., Renner, S.S., 2006. Wax plants disentangled: A phylogeny of *Hoya* (Marsdenieae, Apocynaceae) inferred from nuclear and chloroplast DNA sequences. Molecular Phylogenetics and Evolution 39, 722-733.

Hoya anulata Schltr. in Fl. Schutzgeb.Südsee, 362. 1905.

环冠球兰

分布：印度尼西亚（伊里安查亚）、巴布亚新几内亚和澳大利亚（昆士兰）。

形态特征：攀援或悬垂藤本。茎直径约 3mm。叶片肉质，椭圆形，长约 8cm，宽约 5cm，通常淡绿色，在全日照下边缘或全部变红，叶脉不明显。伞形花序呈平头状，具花 10~15 朵；花直径约 1.5cm，具椰香气；花冠平展，白色，被短茸毛；副花冠淡粉色至粉紫色，中间颜色较深。

栽培习性：喜半阴；喜温暖潮湿的环境；扦插枝条易成活；植株生长快，但攀附性不佳，需适当扶持；本种营养生长期较长，需待植株生长健壮后才会萌生花序。

光照需求：

养护难度：

Distribution: Indonesia (Irian Jaya), Papua New Guinea and Australia (Queensland).

Morphology: Climbing or hanging vine. Stem ca. 3mm diameter. Leaf blade fleshy, elliptic, ca. 8cm long and ca. 5cm wide, pale green, margin or whole leaf turns red in full sunlight, veins obscure. Umbel flat-topped, 10-15 flowered; flower ca. 1.5cm diameter, coconut fragrance; corolla flat, white, puberulent; corona light pink to pale lilac, center darker.

Habits: Grows well in half-shading, warm and moist environment; cuttings have high survival rate; fast-growing, not good at climbing, proper support needed; long vegetative period need to grow healthily and blooms.

Lighting requirement:

Maintenance difficulty:

Hoya archboldiana C. Norman in Brittonia, **2**: 328. 1937.

大花球兰

分布：印度尼西亚和巴布亚新几内亚。

形态特征：攀援藤本。茎直径约 4mm。叶片肉质，长椭圆形至披针形，长约 15cm，宽 4~5cm，新叶淡紫红色，后变为深绿色，中脉明显。伞形花序具花 5~15 朵；花大，直径 4~5cm，具淡栀子花香；花冠阔钟状，裂片向背面反折，黄白色、粉红色至紫红色，里面中间颜色较浅；副花冠颜色较花冠的颜色深，为深红至紫红色。

栽培习性：喜半阴至略明亮的散射光；喜温暖、潮湿的环境；扦插枝条易成活；营养生长期较长，需待植株生长健壮后才会开花，但首次开花后则较常开花；栽培需注意勿将嫩茎向下弯曲，否则茎尖容易干枯。

备注：本种以美国生物学家及探险家 Richard Archbold 命名。

光照需求：

养护难度：

Distribution: Indonesia and Papua New Guinea.

Morphology: Climbing vine. Stem ca. 4mm diameter. Leaf blade fleshy, oblong to lanceolate, 15cm long and 4-5cm wide, young leaf blade bright purple-red, then turns deep green, midvein distinct. Umbel 5-15 flowered; flower large, 4-5cm diameter, light gardenia fragrance; corolla wide campanulate, corolla lobes reflexed, yellowish white, pink to fuchsia, inner center lighter; corona's color deeper than corolla, deep red to fuchsia.

Habits: This *Hoya* likes half-shading to bright scattered light, and grows well in warm and moist environment; cuttings have high survival rate; needs long vegetative period to grow healthily and bloom. After first blooming, bloom often. Do not bend the young stem downward when cultivating, otherwise stem apex might wither.

Note: Plant named after the American biologist and explorer, Richard Archbold.

Lighting requirement:

Maintenance difficulty:

Hoya ariadna Decne.in Prodr 8: 635. 1844.

阿丽亚娜球兰 艾雷尔球兰

分布：印度尼西亚和巴布亚新几内亚。

形态特征：攀援藤本，茎直径约4mm。叶面片厚革质至肉质，阔卵形，长约8cm，宽4~5cm，基部心形，边缘全缘，先端急尖或钝，深绿色，被短茸毛，中脉明显，侧脉不显。伞形花序具花5~10朵；花大，直径约3cm，无香气；花冠平展，黄色至橙黄色，有时浅粉色；副花冠黄色；雄蕊深红色。

栽培习性：喜半阴；喜温暖潮湿的环境，不耐寒，冬季需保温，否则植株易落叶及干枯；扦插枝条成活率高；幼株不耐水湿，浇水过于频繁则茎及枝条易腐烂，健壮植株可适当增加给水。

光照需求：

养护难度：

Distribution: Indonesia and Papua New Guinea.

Morphology: Climbing vine. Stem ca. 4mm diameter. Leaf blade leathery to fleshy, thick, broad ovate, ca. 8cm long and 4-5cm wide, base heart-shaped, margin entire, apex acute or obtuse, deep green, tomentose, midveins distinct, lateral veins obscure. Umbel 5-10 flowered; flower large, ca. 3cm diameter, odourless. Corolla spreading, yellow to orange, rarely light pink; corona yellow; stamens deep red.

Habits: Grows well in half-shading, warm and moist environment; nonhardy, needs insulation in winter, otherwise easy to defoliate and wither; cuttings have high survival rate; young plant not resistant to wetness, frequent irrigation might lead to rotten stems; healthy plants prefer proper water increment.

Lighting requirement:

Maintenance difficulty:

Hoya australis R. Br. ex J. Traill in Trans. Hort. Soc. London, **7**: 28. 1830.

南方球兰

分布：印度尼西亚（伊里安查亚）、巴布亚新几内亚、斐济、汤加和澳大利亚。

形态特征：攀援藤本。茎直径 2~3mm，常被柔毛。叶片革质至肉质，大小及厚薄变化较大，常为长圆形，长约 11cm，宽约 4cm，中间较厚，向边缘变薄，边缘全缘，先端急尖，绿色，被短茸毛，叶脉明显。伞形花序具花达 40 朵；花直径约 1.5cm，具甜香气；花冠略内弯呈爪状，外面白色，内面中心粉红色至紫红色；副花冠白色。

栽培习性：适应性强；喜半阴；干旱至潮湿的环境均可生长；扦插枝条易成活；植株生长较快；花序分化早，盛花期在秋季，其余时间零星开花；本种叶片形态随生长环境不同变化较大。

光照需求：

养护难度：

Distribution: Indonesia (Irian Jaya), Papua New Guinea, Fiji, Tonga and Australia.

Morphology: Climbing vine. Stem 2-3mm diameter, often puberulent. Leaf blade leathery to fleshy, long ovate, variations in size and thickness, ca. 11cm long and ca. 4cm wide, center thick, margin thin; margin entire, apex acute, green, tomentose, veins distinct. Umbel 40 flowered at most, flower ca. 1.5cm diameter, sweet fragrance. Corolla incurved, claw-like, outside white, inside pink to fuchsia; corona white.

Habits: Grows well in half-shading and dry to moist environment; cuttings have high survival rate; fast-growing; inflorescence emerge early; blooms fully in autumn and fragmentarily in other time; shape of leaf blade changes a lot in different environment.

Lighting requirement:

Maintenance difficulty:

Hoya benguetensis Schltr. in Philipp. J. Sci. 1(Suppl.): 301. 1906.

本格特球兰 本格尔顿球兰

分布：菲律宾（吕宋岛）。

形态特征：攀援藤本。茎直径约2mm。叶片肉质，披针形至椭圆形，长约12cm，宽约5cm，先端急尖，通常绿色，光照较强则变为红色。伞形花序具花10~20朵；花直径约1.5cm，具淡香气；花冠向背面反折，橙红色；副花冠深红色至紫红色。

栽培习性：喜半阴，如光照过强，则叶片容易灼伤；栽培环境需湿度适中，生长季节频繁浇水可促进生长；本种开花较频繁，花落后数周内同一花序梗又可再开花。

备注：本种以其产地菲律宾本格特省（Benguet Province）命名。

光照需求：

养护难度：

Distribution: Philippines (Luzon Island).

Morphology: Climbing vine. Stem ca. 2mm diameter. Leaf blade fleshy, lanceolate to ovate, ca. 12cm long and ca. 5cm wide, apex acute, often green and turns red under strong sun light. Umbel 10-20 flowered; flower ca. 1.5cm diameter, slightly fragrance. Corolla reflexed, orange-colored; corona deep red to fuchsia.

Habits: Grows well in half shading; leaves burnt in too much of sun light; moderate humidity preferred; frequent irrigation can promote plant growth in growing season; blooms frequently and same peduncle will bloom again several weeks within the blossom fall.

Note: This *Hoya* comes from Benguet Province of Philippines and it was named after this place.

Lighting requirement:

Maintenance difficulty:

Hoya bella Hook.Bot. Mag. t. 4402. 1848.

贝拉球兰

分布：中国（云南）、缅甸、老挝和泰国。

形态特征：悬垂藤本。茎直径约2mm，被短绒毛。叶柄极短；叶片披针形，长约3cm，宽1.5cm，中脉明显，侧脉不显。伞形花序呈平头状，其花6~10朵，排列整齐；花直径约1.5cm，具淡香气；花冠平展，白色；副花冠粉红色，半透明。

栽培习性：喜半阴，忌强光照射；喜湿润且通风的环境；栽培宜选择保水且透气的基质；成熟枝条扦插易成活；植株生长快且容易开花；本种株形美观，花色秀雅，是我国产球兰中养护容易且栽培普遍的种类。

光照需求：

养护难度：

Distribution: China (Yunnan), Myanmar, Laos and Thailand.

Morphology: Hanging vine. Stem ca. 2mm diameter, tomentose. Leaf blade lanceolate, petiole short, ca. 3cm long and ca. 1.5cm wide, midvein distinct, lateral vein obscure. Umbel flat-topped, 6 to 10 flowered; flowers neatly arranged, light fragrance, ca. 1.5cm diameter. Corolla spreading, white; corona pink, translucent.

Habits: Grows well in half-shading, moist and ventilated environment; avoid strong sun light; water-retaining and ventilate substrate preferred; mature branch cuttings have high survival rate; fast-growing and easy to bloom; easily maintainable and a widely-grown Chinese species, with nice appearance and beautiful flower.

Lighting requirement:

Maintenance difficulty:

Hoya bicolor Kloppenb.in Fraterna, **16**(1): 1. 2003.

双色球兰

分布：菲律宾。

形态特征：攀援藤本。茎木质化，直径约 6mm。叶片革质，硬，椭圆形，长约 13cm，宽约 5cm，边缘波状，先端具小尖头，深绿色，表面粗糙；伞形花序呈球状，具花 30~40 朵；花直径约 1.2cm，具生姜气味；花冠向背面反折，黄色，边缘红褐色；副花冠白色。

栽培习性：适应性强；喜半阴；喜湿润的环境，频繁浇水可促进生长；较耐寒；成熟枝条扦插成活率高；植株生长较缓慢；本种养护简单，植株花序多，花量大，是极佳的观赏种类。

光照需求：

养护难度：

Distribution: Philippines.

Morphology: Climbing vine. Stem woody, ca. 6mm diameter. Leaf blade leathery, rigid, oval, ca. 13cm long and ca. 5cm wide, margin undulant, apex mucronulate, deep green, with rough surface. Umbel convex, 30-40 flowered; flower ca. 1.2cm diameter, gingery fragrance. Corolla reflexed, yellow, with red brown margin; white corona.

Habits: Well-adapted species, it grows well in half-shading and moist environment; frequent irrigation helps its growth; cold-resistant; mature branch cuttings have high survival rate; slowly-growing; easy to maintain; due to great amount of flowers and ample inflorescence, this species has great ornamental value.

Lighting requirement:

Maintenance difficulty:

Hoya blashernaezii Kloppenb.in Fraterna, 12(1): 9. 1999.

布拉斯球兰（新拟）布拉轩球兰

分布： 菲律宾（卡坦端内斯岛）。

形态特征： 攀援藤本。茎直径约2mm。叶片较薄，条状披针形，长约10cm，宽约2.5cm，先端渐尖，亮绿色，叶脉明显。伞形花序具花20~30朵；花直径约1.5cm，具清香气；花冠钟状，具蜡质光泽，初时白色，后变为淡黄白色；副花冠白色。

栽培习性： 适应性强；喜半阴；喜温暖潮湿的环境；扦插枝条易成活；植株生长快；不耐寒，冬季需保温，否则植株易落叶及干枯；本种开花频繁，花期几全年。

备注： 本种以菲律宾植物标本采集人Blas Hernaez 命名。

光照需求：

养护难度：

Distribution: Philippines (Catanduanes Island).

Morphology: Climbing vine. Stem ca. 2mm diameter. Leaf blade thin, linear lanceolate, ca. 10cm long and ca. 2.5cm wide; apex acuminate, bright green, vein instinct. Umbel 20-30 flowered, flower ca. 1.5cm diameter, fragrance. Corolla campanulate, with waxy luster; flowers are white at early stage, and then turn light yellowish white; white corona.

Habits: Well-adapted and grows well in half-shading, warm and moist environment; cuttings have high survival rate; fast-growing; nonhardy, needs insulation in winter, otherwise tends to defoliate or wither; blooms almost throughout the year.

Note: Named after the Philippine collector, Blas Hernaez.

Lighting requirement:

Maintenance difficulty:

Hoya brevialata Kleijn & Donkelaar in Blumea 46(3): 467. 2001.

短翅球兰

分布：印度尼西亚（苏拉威西岛）。

形态特征：攀援藤本，茎直径约3mm。叶片革质，卵形至椭圆形，长约4cm，宽2~3cm，边缘略向内卷，先端钝，浅绿色，表面光滑，叶脉不显。伞形花序呈平头状，具花15~25朵；花直径约6mm，具清香气；花冠向背面强烈反卷，黄色、粉红色至浅紫色；副花冠黄色，边缘深红色。

栽培习性：喜半阴；喜温暖潮湿的环境，不耐寒，冬季易受寒害，需保温；扦插枝条易成活；植株生长较缓慢；本种健壮植株株形紧密，虽为攀援藤本但也可作悬垂种类栽培，叶片在明亮环境下显灰绿色，较美观。

光照需求：

养护难度：

Distribution: Indonesia (Sulawesi Island).

Morphology: Climbing vine. Stem ca. 3mm diameter. Leaf blade leathery, ovate to ellipse, ca. 4cm long and 2-3cm wide, margin slightly involute, apex obtuse, light green, surface smooth, vein obscure. Umbel flat-topped, 15-25 flowered, flower ca. 6mm diameter, light fragrance; corolla strong revolute, yellow, pink to lilac; corona yellow, with dark red margin.

Habits: Grows well in half-shading, warm and moist environment; nonhardy, needs insulation in winter, otherwise tends to suffer from chilling injury; cuttings have high survival rate; slowly-growing; healthy plant tightly arranged, grown as hanging species; leaf blade appears greyish-green in bright environment, high ornamental value.

Lighting requirement:

Maintenance difficulty:

Hoya buotii Kloppenb.in Fraterna, 15(4): 1. 2002.

波特球兰

分布：菲律宾（吕宋岛和巴拉望岛）。

形态特征：攀援藤本。茎直径约 2mm。叶片薄，阔披针形，长约 10cm，宽约 3.5cm，表面光滑，深绿色，中脉明显，侧脉不明显。伞形花序具花 15~25 朵；花直径约 2.3cm，具持久的清甜香气；花冠平展或略内弯，浅黄色，表面被长绢毛；副花冠暗红色，先端透明。

栽培习性：喜半阴；喜高湿的环境；环境适宜则生长较快且易开花；本种由于叶片较薄，如栽培环境湿度低则易干枯。

备注：本种以菲律宾植物学家 Inocencio E. Buot 博士命名。

光照需求：

养护难度：

Distribution: Philippines (Luzon Island and Palawan Island).

Morphology: Climbing vine. Stem ca. 2mm diameter. Leaf blade thin, wide lanceolate, ca. 10cm long and ca. 3.5cm wide, deep green, surface smooth, midvein distinct, lateral vein obscure. Umbel 15-25 flowered; flower ca. 2.3cm diameter, with lasting sweet fragrance. Corolla spreading or slightly incurved, light yellow, sericeous; corona deep red, apex transparent.

Habits: Grows well in half-shading and highly-humid environment; fast-growing and blooms early in suitable environment; leaf blade thin, it tends to wither in low humidity environment.

Note: Named after Philippine botanist, Dr. Inocencio E. Buot.

Lighting requirement:

Maintenance difficulty:

Hoya burmanica Rolfe in Bull. Misc. Inform. Kew, 1920: 343. 1920.

缅甸球兰

分布：中国和缅甸。

形态特征：悬垂藤本。茎直径约3.5mm。叶柄短，长约3mm；叶片肉质，披针形，长约9cm，宽约2.5cm，边缘全缘，先端渐尖，淡绿色，表面常具银色斑点，叶脉不明显。伞形花序具花5~8朵；花直径0.8~1cm，具香气；花冠平展，亮黄色；副花中间红色，边缘黄色。

栽培习性：喜半阴至稍明亮的散射光，健壮植株给光充足有利于开花；较耐寒，喜凉爽环境；成熟枝条扦插易成活；本种如环境适宜则每节均可生出花序。

光照需求：

养护难度：

Distribution: China and Myanmar.

Morphology: Hanging vine. Stem ca. 3.5mm diameter. Petiole short, ca. 3mm long; leaf blade fleshy, lanceolate, ca. 9cm long and ca. 2.5cm wide; margin entire, apex acute, light green, with silver spots on the surface, vein obscure. Umbel 5-8 flowered; flower 0.8-1cm diameter, fragrance. Corolla spreading, bright yellow; corona has red center and yellow margin.

Habits: Grows in half shade to bright scattered light; healthy plant blooms early in abundant sun light; cold-resistant, cool environment preferred; mature branch cuttings have high survival rate; each node of the plant can bore new inflorescence in suitable environment.

Lighting requirement:

Maintenance difficulty:

Hoya callistophylla T. Green in Fraterna, 13(4): 2. 2000.

美叶球兰（新拟）淡味球兰

分布：加里曼丹岛。

形态特征：攀援藤本。茎直径约 4mm。叶片肉质，椭圆形至阔披针形，长约 13cm，宽约 6cm，绿色，光照较强则变为黄绿色，表面粗糙，叶脉明显，在两面突起，深绿色。伞形花序具花 15~25 朵；花直径约 9mm，具柑橘香气；花冠向背面反折，黄色，先端褐色；副花冠白色。

栽培习性：喜半阴；较耐旱；本种植株基部叶较密集，新萌生的枝条叶稀疏；扦插枝条较难发芽，生长亦缓慢，但成株适应性强，是养护容易且株形美观的种类。

光照需求：☀ ☀ ☀ ☀

养护难度：🔧

Distribution: Kalimantan Island.

Morphology: Climbing vine. Stem ca. 4mm diameter. Leaf blade, ovate to wide lanceolate, ca. 13cm long and ca. 6cm wide; green leaves turn yellowish green in strong sun light; surface rough; vein obvious, bulging on both sides, dark green. Umbel 15-25 flowered; flower ca. 9mm diameter, with cirtus fragrance; corolla reflexed, yellow, apex brown; corona white.

Habits: Grows in half shade; drought-enduring; basal leaves densely; upper leaves loose; cuttings hard to survive and slowly-growing, but mature plant well-adapted, easy-cultivated and beautiful appearance.

Lighting requirement: ☀ ☀ ☀ ☀

Maintenance difficulty: 🔧

Hoya campanulata Blume, Bijdr. 1064, "1826."

钟花球兰 康蓬那球兰

分布：菲律宾、马来西亚和印度尼西亚。

形态特征：攀援藤本。老茎木质化，直径约 3mm。叶片略肉质，椭圆形至长圆形，长约 15cm，宽约 5cm，先端渐尖，深绿色，表面有时具银斑。伞形花序具花 10~15 朵；花直径约 1.8cm，具柠檬香气；花冠阔钟状，白色至淡黄白色，背面有时紫红色；副花冠淡黄白色。

栽培习性：喜阴，光照过强易灼伤叶片；较耐干旱；栽培宜用疏松透气的基质；本种浇水宜湿润叶片，可促进生长。

光照需求：

养护难度：

Distribution: Philippines, Malaysia and Indonesia.

Morphology: Climbing vine. Stem ca. 3mm diameter, turns woody in mature stage. Leaf blade slightly fleshy, ovate to ellipse, ca. 15cm long and ca. 5cm wide; dark green, apex acuminate, sometimes silver spots on the surface. Umbel 10-15 flowered; flower ca. 1.8cm diameter, lemon fragrance; corolla wide campanulate, white to light yellowish white, abaxially sometimes fuchsia; corona light yellowish white.

Habits: Grows in shade environment; leaves burn in too much of sun light; drought-enduring; loose and ventilate substrate preferred; during irrigation the plant, moistening the leaves can help its growing.

Lighting requirement:

Maintenance difficulty:

Hoya camphorifolia Warb.in Fragm. Fl. Philipp. **1**: 129. 1904.

樟叶球兰

分布：菲律宾（吕宋岛）。

形态特征：攀援藤本，茎直径约 2mm；叶片革质，椭圆形，长 7~9cm，宽 3~4cm，边缘全缘，先端渐尖，绿色，叶脉明显。伞形花序呈球状，具花 20~35 朵；花直径约 5mm，具淡香气；花冠平展或内弯呈爪状，粉红色至淡橙红色；副花冠淡紫红色至红色。

栽培习性：适应性强；喜半阴；喜温暖湿润的环境；栽培宜选用疏松、透气的基质；成熟枝条扦插易成活；植株生长快，但嫩枝细弱，需适当扶持其攀爬；本种终年开花不断，花量大，需经常追肥以支持生长。

光照需求：

养护难度：

Distribution: Philippines (Luzon Island).

Morphology: Climbing vine. Stem ca. 4mm diameter. Leaf blade leathery, ellipse, 7-9cm long and 3-4cm wide, margin entire, apex acute, green, vein distinct. Umbel convex, 20-35 flowered; floweer ca. 5mm diameter, light fragrance. Corolla spreading or recurved, pink to light orange; corona light fuchsia to red.

Habits: Well-adapted species, it grows well in half-shading, warm and moist environment; loose and ventilate substrat preferred; mature branch cuttings have high survival rate; fast-growing, young branches are thin and fragile and need proper support; blooms all year with ample flowers, extra fertilizers often need to maintain growing.

Lighting requirement:

Maintenance difficulty:

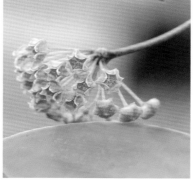

Hoya carnosa (L. f.) R. Br. in Mem. Wern. Nat. Hist. Soc. **1**: 27. 1811[1810].

球兰

分布：中国（长江以南）、日本、印度、越南和马来西亚。

形态特征：攀援藤本。茎直径约 2mm。叶片肉质，椭圆形，先端尖，大小变化较大，深绿色，有时具灰色斑点。伞形花序球形，直径 8~10cm，具花 20~40 朵；花直径 1.6~2cm，具浓郁香气，夜间尤甚；花冠平展，白色、粉红色至深红色，被短茸毛；副花冠中间深红色，边缘白色至淡黄白色。

栽培习性：喜半阴至明亮环境，可露天栽培，光照充足则开花较多；较耐寒，可耐 5℃ 左右的低温；扦插枝条易生根；幼株需较长的生长期，待健壮后才会萌生花序；花期可达 15 天，花凋落数周后同一花序梗会再度开花。

光照需求：

养护难度：

Distribution: China (the southern region of Yangtze River), Japan, India, Vietnam and Malaysia.

Morphology: Climbing vine. Stem ca. 2mm diameter. Leaf blade fleshy, ellipse, apex accuminate, various in sizes, dark green, sometimes with grey spots. Umbel convex, 20-40 flowered; flower 1.6-2cm diameter, strong fragrance, becomes stronger at night. Corolla spreading, white, pink to dark red, tomentose; corona center dark red, margin white to light yellowish white.

Habits: Grows well in half shade to bright environment and also grown outdoor; blooms ample flowers in sufficient sun light; species can endure a low temperature (5℃); cuttings have high survival rate; young plant need long growing period, and bore inflorescence after plant maturation; florescence can last 15 days; same peduncle bloom again several weeks after the flowers wither.

Lighting requirement:

Maintenance difficulty:

Hoya caudata J. D. Hook. Fl. Brit. India. 4(10): 60. 1883.

尾状球兰 青铜器球兰

分布：泰国（南部）和马来西亚。

形态特征：攀援藤本。茎直径 3mm。新叶柔软，成熟后肉质，易断裂，椭圆状披针形，长约 9cm，宽约 5cm，边缘有时波状，先端尾尖，下面常红色，上面深绿色，光照较强则变为红色，两面密布银色斑块。伞形花序具花 8~10 朵；花直径约 2cm，无香气或有淡香气；花冠平展，先端向背面反卷，白色、淡黄白色至浅黄色，被白色长柔毛；副花冠红色，边缘淡棕色；雄蕊先端具 5 条毛状附属物。

栽培习性：喜半阴；栽培宜用疏松多隙的混合基质；扦插枝条易成活；植株初期生长较慢，后期加快；需较长生长期，待生长健壮后才会萌生花序。

光照需求：☀☀☀

养护难度：🌱🌱

Distribution: Thailand (southern region) and Malaysia.

Morphology: Climbing vine. Stem ca. 3mm diameter. Young leaves soft and turn fleshy and crispwhen mature, ellipse-lanceolate, ca. 9cm long and ca. 5cm wide; margin sometimes undulant, apex caudate, abaxially often red, adaxially dark green or turns red in strong sun light, with dense silver spots on both sides. Umbel 8-10 flowered; flower ca. 2cm diameter, odourless or light fragrance; corolla spreading, apex revolute, white, light yellowish white to light yellow, puberulent; corona red, with brown margin; five hairy appurtenances on the top of each stamen.

Habits: Grows well in half shade; loose and porous substrate mixture is preferred; cuttings have high survival rate; slowly-growing at early stage, later grows faster; long growing period; the plant bore inflorescence after maturation.

Lighting requirement: ☀☀☀

Maintenance difficulty: 🌱🌱

Hoya chinghungensis (Tsiang & P. T. Li) M. G. Gilbert, P. T. Li & W. D. Stevens in Novon, **5**(1): 9. 1995.

景洪球兰

分布：中国（云南）。

形态特征：悬垂藤本。茎直径约 3mm，被短茸毛。叶片肉质，心形，长约 1.5cm，宽约 1cm，边缘全缘，先端钝，绿色，表面被短茸毛。伞形花序平头状，具花 5~6 朵，排列整齐；花直径约 1.3cm，有淡香气；花冠平展，先端向背面反卷，白色；副花冠红色，中间黄色，透明。

栽培习性：适应性强，易养护，在半阴环境下栽培生长较快且叶片具光泽；栽培宜用疏松透气的基质，忌积水；扦插宜采用成熟枝条；本种是我国产球兰种类中美观且栽培较多的种类。

备注：本种以其产地中国云南景洪市命名。

光照需求：

养护难度：

Distribution: China (Yunnan).

Morphology: Hanging vine. Stem ca. 3mm diameter, tomentose. Leaf blade fleshy, heart-shaped, ca. 1.5cm long and ca. 1cm wide; margin entire, apex obtuse; green, tomentose. Umbel flat-topped, 5-6 flowered, flowers neatly arranged, ca. 1.3cm diameter, light fragrance; corolla spreading, revolute, white; corona red, center yellow and transparent.

Habits: Well-adapted and maintainable; plant grows fast in half-shading with lustrous leaf blades; loose and ventilate substrate preferred, avoid water logging; mature branches better for cutting; beautiful appearance and widely cultivated in China.

Note: This *Hoya* comes from Jinghong City of Yunnan, China and it was Named after this place.

Lighting requirement:

Maintenance difficulty:

Hoya chlorantha Rech.in Repert. Spec. Nov. Regni Veg. 5: 131. 1908.

绿花球兰

分布：萨摩亚群岛。

形态特征：攀援藤本。茎直径约 2mm。叶片略肉质，椭圆体形至倒披针形，长约 10cm，宽 1.5~2cm，绿色，叶脉明显。伞形花序具花 12~18 朵；花直径约 1.5cm，几无香气；花冠平展或略内弯，先端向背面反卷，淡黄绿色，表面被短绒毛；副花冠黄色，中间有时具红晕。

栽培习性：适应性强，对生长环境要求不高；喜半阴；不耐寒，冬季栽培需控水，8℃以下易受寒害；扦插枝条易生根，环境适宜则生长快。

光照需求：
养护难度：

Distribution: Samoa Islands.

Morphology: Climbing vine. Stem ca. 2mm diameter. Leaf blade fleshy, ellipse to oblanceolate, ca. 10cm long and 1.5-2cm wide, green, vein distinct. Umbel 2-18 flowered; flower ca. 1.5cm diameter, nearly odourless; corolla spreading or slightly incurved, apex revolute, light yellow green, tomentose; corona yellow, sometimes with red center.

Habits: Well-adapted and low demand in growing environment; plant likes half shade; nonhardy, water controlling needed in winter; tends to suffer from chilling injury when the temperature falls down below 8 ℃ ; cuttings have high survival rate; fast-growing in suitable environment.

Lighting requirement:
Maintenance difficulty:

Hoya chunii P. T. Li in Bull. Bot. Lab. N.-E. Forest. Inst., Harbin, 4(1): 121. 1984.

椰香球兰

分布：巴布亚新几内亚。

形态特征：攀援藤本。茎直径约 3mm。叶片肉质，椭圆形至阔椭圆形，长约 12cm，宽约 8cm，边缘全缘或波状，先端尖，浅绿色，叶脉明显，伞形花序呈球状，具花约 30 朵；花直径约 9mm，具甜香气；花冠向背面反卷，淡粉色至杏黄色；副花冠粉红色至白色。

栽培习性：适应性强；喜光照充足的环境，如在荫蔽环境下则叶片易徒长并变形；扦插枝条易成活，生长较快；本种营养生长期较长，需栽培一定时间后才萌生花序。

备注：本种以我国著名植物学家陈焕镛（Chun Woon-Young）命名。

光照需求：

养护难度：

Distribution: Papua New Guinea.

Morphology: Climbing vine. Stem ca. 3mm diameter. Leaf blade fleshy, ellipse to wide ellipse, ca. 12cm long and ca. 8cm wide, margin entire or undulant, apex acute; light green, vein distinct. Umbel convex, ca. 30 flowered; flower ca. 9mm diameter, sweet fragrance; corolla revolute, light pink to apricot; corona pink to white.

Habits: Well-adapted and likes full sun light; leaf blade tends to overgrow and deform in shading environment; cuttings have high survival rate and grow fast; needs a long period to emerge inflorescence.

Note: Named after the famous Chinese botanist, Chun Woon Young.

Lighting requirement:

Maintenance difficulty:

Hoya ciliata Teijsm.& Binn.in Cat. Hort. Bot. Bogor. 129. 1866.

纤毛球兰

分布：菲律宾。

形态特征：攀援藤本。茎直径约 3cm。叶片椭圆形，长约 8cm，宽约 5cm，边缘全缘，先端具小尖头，绿色至淡绿色，表面被短茸毛，叶脉不明显。伞形花序具花 3~8 朵，花直径可达 4cm，几无香气；花冠常向背面反折，红棕色；副花冠中间红色，边缘黄色。

栽培习性：喜半阴至明亮的散射光，光线不足则难开花；栽培环境湿度不宜过高；浇水不宜过于频繁，冬季需控水；宜选用疏松透气的栽培基质；成熟枝条扦插成活率更高；环境适宜一年可多次开花。

光照需求：

养护难度：

Distribution: Philippines.

Morphology: Climbing vine. Stem ca. 3mm diameter. Leaf blade ellipse, ca. 8cm long and ca. 5cm wide, margin entire, apex mucronulate, light green to green, tomentose, vein obscure. Umbel 3-8 flowered; flower ca. 4cm diameter, nearly odourless; corolla reflexed, red brown; corona center red, margin yellow.

Habits: Grows well in half shade to bright scattered light; hardly blooms in deficient sun light; moderate humidity preferred; avoid frequent irrigation and water controlling needed in cold season; loose and ventilate substrate preferred; mature cuttings have high survival rate; blooms several times throughout the year in suitable environment.

Lighting requirement:

Maintenance difficulty:

Hoya cinnamomifolia Hook. in Bot. Mag. 4347. 1848.

桂叶球兰（新拟）玉桂球兰

分布： 印度尼西亚（爪哇岛）。

形态特征： 攀援藤本，茎直径约4mm，叶片大，略肉质，椭圆形至长卵形，长11~15cm，宽4~6cm，基部截形，边缘全缘，先端渐尖至急尖，绿色至深绿色，叶脉明显。伞形花序呈球状，具花15~24朵；花直径约1.6cm，几无香气。花冠向背面反折，浅绿色至黄绿色；副花冠深红色至紫红色。

栽培习性： 喜阴；喜凉爽、通风的环境，较耐寒；栽培宜选用疏松、透气的基质；扦插枝条成活率高；植株生长需肥较多，可适当增加施肥；本种花色特别，是独特的观赏种类。

光照需求：
养护难度：

Distribution: Indonesia (Java Island).

Morphology: Climbing vine. Stem ca. 4mm diameter; Leaf blade large, slightly fleshy, ellipse or oblong, 11-15cm long and 4-6cm wide, base truncate, margin entire, apex acuminate to acute, green to deep green, vein distinct. Umbel 15-24 flowered; flower ca. 1.6cm diameter, nearly odourless. Corolla reflexed, light green to yellow green; corona dark red to fuchsia.

Habits: Grows well in shading, cool and ventilate environment; cold-resistant; loose and ventilate substrate preferred; cuttings have high survival rate; high demand in fertilization, proper addition of fertilizers needed; special flower colors prove its high ornamental value.

Lighting requirement:
Maintenance difficulty:

Hoya collina Schltr. in Bot. Jahrb. Syst. 50: 111. 1913.

丘生球兰（新拟） 小丘球兰

分布： 印度尼西亚（比亚克岛）和巴布亚新几内亚。

形态特征： 攀援藤本，茎直径约 4mm。叶片肉质，椭圆形，长约 5cm，宽约 3cm，边缘全缘，先端急尖或钝，绿色，有时部分或全部变为橙红色至红色，表面光滑，叶脉不显。伞形花序具花 4~10 朵；花直径约 1cm，具清香气；花冠平展，黄色，被短茸毛；副花冠黄色，边缘常红色。

栽培习性： 喜明亮的散射光；喜温暖、潮湿的环境，环境湿度高则可萌生较多的气生根，不耐寒，冬季受寒叶片易脱落；扦插枝条易成活；植株生长快，但攀援性不强，需适当扶持或使其悬垂生长。

光照需求：
养护难度：

Distribution: Indonesia (Biak Island) and Papua New Guinea.

Morphology: Climbing vine. Stem ca. 4mm diameter. Leaf blade fleshy, ellipse, ca. 5cm long and ca. 3cm wide, margin entire, apex acute or obtuse, green, sometimes leaf blade turns partially or entirely orange to red, surface smooth, vein obscure. Umbel 4-10 flowered; flower ca. 1cm diameter, fragrance; corolla spreading, yellow, tomentose; corona yellow, often with red margin.

Habits: Grows well in bright scattered light; warm and moist environment preferred; take more aerial roots in high humidity environment; nonhardy, leaves tend to fall when suffer chilling injuryin winter; cuttings have high survival rate; fast-growing plant, but not good at climbing, proper support needed.

Lighting requirement:
Maintenance difficulty:

Hoya coriacea Blume, Bijdr. 1061. 1826.

革叶球兰

分布：泰国、菲律宾、马来西亚和印度尼西亚。

形态特征：攀援藤本。茎直径约 2mm。叶片较薄，椭圆形，长约 12cm，宽 5~6cm，边缘全缘，先端尾尖，深绿色，叶脉明显。伞形花序呈球形，具花 30~40 朵；花直径约 2cm，具柑橘香气；花冠向背面反折，先端向背部反卷，淡黄色，表面被短茸毛；副花冠白色，中间有时紫色。

栽培习性：喜半阴至光照充足的环境；不耐干旱，但栽培基质也不宜过于湿润；扦插枝条易成活，生长较快；本种植株紧凑，常年开花，花序较大，花气味清香，是优良的观赏种类。

光照需求：☀☀☀

养护难度：🌱🌱

Distribution: Thailand, Philippines, Malaysia and Indonesia.

Morphology: Climbing vine. Stem ca. 2mm diameter. Leaf blade thin, ellipse, ca. 12cm long and 5-6cm wide, margin entire, apex caudate, deep green, vein distinct. Umbel convex, 30-40 flowered; flower ca. 2cm diameter, citrus fragrance; corolla reflexed, apex revolute, light yellow, tomentose; corona white, sometimes with purple center.

Habits: Plant grows well in half shade to sufficient light; drought-sensitive, but substrate should not be too moist; cuttings have high survival rate; fast growing; compact with thick leaves; blooms throughout the year; large inflorescence with fragrance, having great ornamental value.

Lighting requirement: ☀☀☀

Maintenance difficulty: 🌱🌱

Hoya coronaria Blume, Bijdr. Fl. Ned. Ind. 1063. 1826.

冠花球兰（新拟） 冠状球兰

分布：马来西亚。

形态特征：攀援藤本，茎肉质，直径约5mm。叶片肉质，椭圆形，长约7cm，宽约4cm，边缘全缘，先端急尖，绿色，被短茸毛，中脉明显，侧脉不明显。伞形花序具花3~5朵；花直径约3cm，无香气；花冠肉质，平展，白色、粉红色至粉紫色；副花冠中间粉红色至粉紫色，边缘黄白色。

栽培习性：喜明亮的散射光；喜温暖、潮湿且通风的环境，不耐水湿，肉质茎干遇积水易患黑腐病，不耐寒，冬季需适当保温；栽培宜选用疏松、透气且不积水的基质；如环境适宜则植株生长迅速，株形较大，需预留足够空间供其生长。

光照需求：★★★★★

养护难度：🌱🌱

Distribution: Malaysia.

Morphology: Climbing vine. Stem fleshy, ca. 5mm diameter. Leaf blade fleshy, ellipse, ca. 7cm long and ca. 4cm wide, margin entire, apex acute, green, tomentose, midvein distinct, lateral vein obscure. Umbel 3-5 flowered; flower ca. 3cm diameter, odourless; corolla fleshy, spreading, white, pink to pale lilac; corona center pink to pale lilac, margin yellowish white.

Habits: Plant likes bright scattered light, and prefers warm, moist and ventilate environment; not humidity-resistant, fleshy stem tends to suffer from black rot in water logging; nonhardy, needs proper insulation in cold seasons; loose, ventilate and nonwater-logged substrate preferred; fast-growing in suitable environment; due to the large size of the plant, large growing space should be needed.

Lighting requirement: ★★★★★

Maintenance difficulty: 🌱🌱

Hoya cumingiana Decne.in Prodr. 8: 636. 1844.

卡氏球兰（新拟） 孜然球兰

分布： 菲律宾（吕宋岛、巴拉望岛和民都洛岛）和印度尼西亚（爪哇岛）。

形态特征： 攀援藤本，枝条常披散。茎直径 2~3mm。叶片椭圆形，长约 3cm，宽约 2cm，淡绿色至绿色，叶脉不明显。伞形花序具花 10~20 朵，花谢后花序梗一同脱落；花直径 0.8~1cm，具柑橘香气；花冠淡黄绿色，向背面反折；副花冠紫色。

栽培习性： 适应性强，能适应不同的栽培环境和基质；扦插枝条易成活；生长快；稍强的光照能使叶片更肥厚，从而能使植株更耐低温。

备注： 本种以英国植物标本采集人 Hugh Cuming 命名。

光照需求： ●●●
养护难度： 🌱

Distribution: Philippines (Luzon Island, Palawan Island and Mindoro Island) and Indonesia (Java Island).

Morphology: Climbing vine, often with hanging down branches. Stem 2-4mm diameter. Leaf blade ellipse, ca. 3cm long and ca. 2cm wide, light green to green, vein obscure. Umbel 10-20 flowered; flower 0.8-1cm diameter, citrus fragrance; peduncle drops when flowers wither; corolla reflexed, light yellow green; corona purple.

Habits: Well-adapted, accommodate to different substrates and environment; cuttings have high survival rate; fast-growing; strong light may produce more fleshy leaves, and make the plant more cold-resistant.

Note: Named after the British plant collector, Hugh Cuming.

Lighting requirement: ●●●
Maintenance difficulty: 🌱

Hoya danumensis Rodda & Nyhuus in Webbia, 64(2): 164 (-165; fig. 1), 2009.

丹侬球兰（新拟） 达奴姆球兰

分布：马来西亚（沙巴）。

形态特征：攀援灌木。茎直径3~5mm，老茎常木质化。叶片略肉质，椭圆形，长6~10cm，宽4~5cm，先端具尖头，较薄，深绿色，叶脉明显。伞形花序具花12~18朵；花直径约2.5cm，具柑橘香气；花冠阔钟状，白色至粉红色；副花冠直径约6mm，中心白色至粉红色，边缘淡黄白色。

栽培习性：喜半阴；喜高温和高湿的环境；扦插枝条较难生根，宜选用稍木质化的枝条；成活后植株生长较快且花序分化较早；本种花大且颜色素雅，是较特别的种类。

备注：本种以其产地马来西亚丹侬谷保护区（Danum Valley Conservation Area）命名。

光照需求：☀ ☀ ☀

养护难度：🌱 🌱

Distribution: Malaysia (Sabah).

Morphology: Climbing shrub. Stem 3-5mm diameter; matured stems always lignified. Leaf blade fleshy, ellipse, 6-10cm long and 4-5cm wide, apex mucronulate, thin, deep green, vein distinct. Umbel 12-18 flowered; flower ca. 2.5cm diameter, citrus fragrance; corolla wide campanulate, white to pink; corona ca. 6mm diameter, center white to pink, margin yellowish white.

Habits: Grows well in half shade, high temperature and high humidity; cuttings hardly form roots; slight lignified branch cuttings preferred; fast-growing, inflorescence emerge early; large and elegant flowers make the species very special.

Note: This *Hoya* comes from Danum Valle Conservation Area, Malaysia and it was named after this place.

Lighting requirement: ☀ ☀ ☀

Maintenance difficulty: 🌱 🌱

Hoya darwinii Loher in Gard. Chron. **47**: 66. 1910.

达尔文球兰（新拟） 多温球兰 蚁球球兰

分布：菲律宾（吕宋岛）。

形态特征：攀援藤本。茎直径约 3mm。叶片革质，椭圆形至披针形，长约 10cm，宽 4~5cm，边缘全缘，先端渐尖，浅绿色至绿色，中脉明显，侧脉不显。伞形花序具花 10~20 朵；花直径约 1.8cm，无香气；花冠向背面反折，白色至粉紫色；副花冠中间红色至紫红色，边缘白色。

栽培习性：适应性强；喜半阴；喜温暖潮湿的环境；栽培宜选用疏松、透气的基质，忌积水；扦插枝条易生根；植株生长快，易开花，可适当增加施肥；本种的特点是叶片受外界刺激（如叮咬或啃食）则卷曲呈球状。

备注：本种以英国著名生物学家 Charles R. Darwin 命名。

光照需求：☀ ☀ ☀

养护难度：🌱

Distribution: Philippines (Luzon Island).

Morphology: Climbing vine. Stem ca. 3mm diameter. Leaf blade leathery, ellipse to lanceolate, ca. 10cm long and 4-5cm wide, margin entire, apex acuminate, light green to green, midvein distinct, lateral vein obscure. Umbel 10-20 flowered; flower ca. 1.8cm diameter, odourless; corolla reflexed, white to pale lilac; corona center red to fuchsia, corona margin white.

Habits: Well-adapted and grows well in half-shading, warm and moist environment; loose and ventilate substrate preferred; avoid water logging; cuttings have high survival rate; plant grow fast and blooms early; proper increment in fertilizers needed; leaf blade frizzles to ball shape under external stimulations (such as biting or nibbling).

Note: Named after the famous British biologist, Charles R. Darwin.

Lighting requirement: ☀ ☀ ☀

Maintenance difficulty: 🌱

Hoya dasyantha Tsiang in Sunyatsenia, **3**(2–3): 174–175. 1936.

厚花球兰

分布：中国和越南。

形态特征：攀援藤本。茎直径约 3mm。叶片椭圆形至阔披针形，长约 10cm，宽约 5cm，边缘全缘，先端渐尖，绿色，叶脉较明显。伞形花序具花 15~20 朵；花直径约 1.5cm，具芳香气；花冠平展，先端向背部反卷，花冠白色至淡粉红色；副花冠中心红色，边缘白色至淡黄白色。本种形态与护耳草 *Hoya fungii* 较相似，但花较小。

栽培习性：适应性强，在露天或半阴的环境下栽培均生长良好；较耐寒；幼嫩或成熟的枝条扦插均易成活。

光照需求：

养护难度：

Distribution: China and Vietnam.

Morphology: Climbing vine. Stem ca. 3mm diameter. Leaf blade ellipse to wide lanceolate, ca. 10cm long and ca. 5cm wide, margin entire, apex acuminate, green, vein distinct. Umbel 15-20 flowered; flower ca. 1.5cm diameter, fragrance; corolla spreading, apex revolute, white to pink; corona center red, margin white to yellowish white. This species is similar to *Hoya fungii*, but with smaller flowers.

Habits: Well-adapted and likes half shade or open-air both environment; cold-resistant; young and mature cuttings have high survival rate.

Lighting requirement:

Maintenance difficulty:

Hoya davidcummingii Kloppenb. in Fraterna, **1995**(2): 10–11. 1995.

戴维德球兰

分布：菲律宾。

形态特征：悬垂藤本。茎直径约 2mm。叶片肉质，披针形，长约 6cm，宽约 2cm，先端渐尖，上面绿色，下面深绿色，边缘常具墨绿色边。伞形花序具花 10~15 朵，花直径约 7mm，具甜香气；花冠浅钟状，边缘常向背面反卷，稀平展，粉红色至淡紫红色，表面被短茸毛；副花冠中心红色，边缘黄色。

栽培习性：适应性强；喜阴，适当遮阴则生长较佳；喜通风的环境；栽培宜选择透气的基质，忌积水，否则叶片易腐烂；不耐寒，冬季需适当保温；选用成熟枝条扦插生根较快。

备注：本种以植物学家 David Cumming 命名。

光照需求：

养护难度：

Distribution: Philippines.

Morphology: Hanging vine. Stem ca. 2mm diameter. Leaf blade fleshy, lanceolate, ca. 6cm long and ca. 2cm wide, apex acuminate, abaxially green, adaxially deep green, margin blackish green. Umbel 10-15 flowered; flower ca. 7mm diameter, sweet fragrance; corolla shallow campanulate, margin revolute, rare spreading, pink to light fuchsia, tomentose; corona center red, margin yellow.

Habits: Well-adapted and grows well in properly shading space and ventilate environment; ventilate substrate preferred; avoid water logging, leaf blade tends to rot; nonhardy, needs proper insulation during cold seasons; mature branch cuttings form roots easily.

Note: Named after the botanist, David Cumming.

Lighting requirement:

Maintenance difficulty:

Hoya densifolia Turcz. in Bull. Soc. Imp. Naturalistes Moscou, **21**(1): 261. 1848.

密叶球兰

分布：菲律宾和印度尼西亚（爪哇岛）。

形态特征：半灌木。茎直立或缠绕，直径约 4mm。叶片厚纸质，椭圆形，长约 6cm，宽约 3cm，边缘全缘，先端急尖，淡绿色，新叶表面具微柔毛，后变无毛。伞形花序生于茎先端，具花 10~20 朵，开花后有时花序梗随花脱落；花直径约 1.2cm，具柑橘香气；花冠向背面反折，淡黄绿色至淡绿色；副花冠紫红色。

栽培习性：适应性强；喜半阴；较耐寒冷，但在温暖潮湿的环境生长更佳；扦插枝条易成活；本种花序着生于枝条先端，常多个花序同时开花，花期长达 10 天左右，成株可适当摘心以促进侧枝生长，开花则更为繁茂。

光照需求：☀ ☀ ☀

养护难度：🛠

Distribution: Philippines and Indonesia (Java Island).

Morphology: Subshrub; stem erect or twining, ca. 4mm diameter. Leaf blade thick papyraceous, ellipse, ca. 6cm long and ca. 3cm wide, margin entire, apex acute, light green, new leaves puberulent, then become glabrous. Umbel on the upper part of branch, sometimes puduncle drops when flowers wither; 10-20 flowered; flower ca. 1.2cm diameter, citrus fragrance; corolla reflexed, light yellowish green to light green; corona fuchsia.

Habits: Well-adapted and likes half shade; cold-resistant, warm and moist environment preferred; inflorescence on the branch apex; inflorescences bloom at the same time, which can last about 10 days; proper top removal can help to grow lateral branches and the blossom will be more lush.

Lighting requirement: ☀ ☀ ☀

Maintenance difficulty: 🛠

Hoya deykeae T. Green in Fraterna, **13**(1): 15. 2000 ["deykei"].

戴克球兰 德克球兰 芭蕉扇球兰

分布：印度尼西亚（苏门答腊岛）。

形态特征：缠绕藤本。茎直径约 3mm，略木质化。叶片肉质，倒心形，长约 10cm，宽约 6cm，亮绿色至红色，表面有时具银色斑块，叶脉明显。伞形花序具花 20~35 朵；花直径约 6mm，具柑橘香气；花冠全部向背面反卷，乳白色至浅黄色；副花冠白色至淡黄白色。

栽培习性：喜明亮的散射光，置于明亮处则叶片变为红色，较美观；不耐寒，冬季需适当保温；较易患黑斑病；扦插宜选用成熟枝条；植株生长较缓慢。

光照需求：

养护难度：

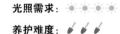

Distribution: Indonesia (Sumatra Island).

Morphology: Twining vine. Stem slightly woody, ca. 3mm diameter. Leaf blade fleshy, obcordate, ca. 10cm long and ca. 6cm wide, bright green to red, vein distinct, sometimes silver spots on the surface. Umbel 20-35 flowered; flower ca. 6mm diameter, citrus fragrance; corolla fully revolute, milk white to light yellow; corona white to light yellowish white.

Habits: Grows well in bright scattered light; leaf blade turns to red in bright place; nonhardy, needs proper insulation duringcold season; plant tends to suffer from black spot disease; better to use mature branches as cuttings; slowly-growing.

Lighting requirement:

Maintenance difficulty:

Hoya diversifolia Blume, Bijdr. Fl. Ned. Ind. 1064. 1826.

异叶球兰

分布：中国（海南）、缅甸、越南、老挝、柬埔寨、菲律宾、泰国、马来西亚（婆罗洲岛）和印度尼西亚（加里曼丹岛、苏门答腊岛和爪哇岛）。

形态特征：攀援藤本，茎直径约 4mm。叶片肉质，椭圆形，长约 9cm，宽约 4cm，先端尖，淡绿色。伞形花序呈球形，具花 10~18 朵；花直径 1.4cm，无香气；花冠平展，厚实，粉红色，被短茸毛；副花冠粉红色。

栽培习性：适应性强，露天或半阴环境栽培均可，光照充足则开花繁茂；较耐寒，冬季可耐受 5℃ 左右低温；扦插枝条易成活；生长较快，但营养生长期较长，需待植株生长健壮后才会萌生花序。

光照需求：☀☀☀☀☀

养护难度：🌱

Distribution: China (Hainan), Myanmar, Vietnam, Laos, Cambodia, Philippines, Thailand, Malaysia (Borneo Island) and Indonesia (Kalimantan Island, Sumatra Island and Java Island).

Morphology: Climbing vine. Stem ca. 4mm diameter. Leaf blade fleshy, ellipse, ca. 9cm long and ca. 4cm wide, apex sharp, light green. Umbel convex, 10-18 flowered; flower ca. 1.4cm diameter, odourless; corolla spreading, thick and solid, pink, tomentose; corona pink.

Habits: Well-adapted and grows well in open place and half-shading environment; blooms lushly in sufficient sun light; cold-resistant, endure in low temperature (5℃) during cold seasons; cuttings have high survival rate; fast-growing; needs a long period to give blossom after plant grows maturation.

Lighting requirement: ☀☀☀☀☀

Maintenance difficulty: 🌱

Hoya elliptica J. D. Hook. Fl. Brit. India, 4(10): 58. 1883.

椭圆叶球兰

分布：泰国和马来西亚。

形态特征：缠绕藤本。茎直径约 4mm。叶片略肉质，椭圆形，长约 8cm，宽约 5cm，淡绿色，有时具银色斑点，中脉明显，侧脉清晰。伞形花序具花 20~35 朵；花直径约 1.4cm，具淡香气；花冠平展或略内弯，花冠白色，先端淡黄白色，表面具微柔毛；副花冠中间红色，边缘白色。

栽培习性：喜光；喜通风且湿润的环境，如闷热则极易落叶；不耐寒冷；栽培宜选择透气的基质，忌积水；扦插宜选择成熟的枝条；条件适宜则生长极快；攀附能力不强，需适当扶持。

光照需求：☀ ☀ ☀ ☀ ☀

养护难度：🌱 🌱

Distribution: Thailand and Malaysia.

Morphology: Twining vine. Stem ca. 4mm diameter. Leaf blade fleshy, ellipse, ca. 8cm long and ca. 5cm wide, light green, sometimes with silver spots, midvein and lateral vein distinct. Umbel 20-35 flowered; flower ca. 1.4cm diameter, light fragrance; corolla spreading or recurved, white, corolla apex light yellowish white, puberulent; corona center red, margin white.

Habits: Plant likes light, ventilate and moist environment; leaves tend to drop in sultry space; nonhardy; ventilate substrate preferred; avoid water logging; mature branches are better for cuttings; fast-growing in proper living conditions; not good at climbing, so proper support needed.

Lighting requirement: ☀ ☀ ☀ ☀ ☀

Maintenance difficulty: 🌱 🌱

Hoya engleriana Hosseus in Notizbl. Bot. Gart. Berlin–Dahlem, 4: 315. 1907.

恩格勒球兰

分布：印度、不丹、尼泊尔、越南、柬埔寨和泰国（北部）。

形态特征：悬垂藤本。茎直径 1~2mm，被短茸毛。叶三枚轮生；叶片略肉质，披针形，长约 2.5cm，宽约 5mm，边缘向背面反卷，先端钝，绿色，表面粗糙，中脉在背面明显。伞形花序平头状，具花 5~7 朵，花排列整齐，花序梗长约 6mm，开花后常随花一同脱落；花直径约 1.8cm，具淡香气；花冠白色，平展；副花冠粉红色，略透明。

栽培习性：喜半阴，光照太强则叶片生长较差；喜温暖潮湿的环境；较耐寒，适当降温可促进开花；栽培宜选用疏松、透气的基质；扦插适宜选用成熟枝条，成活后生长较快。

备注：本种以德国著名植物学家 Adolph Engler 命名。

光照需求：
养护难度：

Distribution: India, Bhutan, Nepal, Vietnam, Cambodia and Thailand (northern region).

Morphology: Hanging vine. Stem 1-2mm diameter, tomentose. Leaf 3 verticillate; leaf blade fleshy, lanceolate, ca. 2.5cm long and ca. 5mm wide, margin revolute, apex obtuse, green, surface rough, abaxially midvein distinct. Umbel flat-topped, 5-7 flowered; flower ca. 1.8cm diameter, light fragrance; flowers neatly arranged; peduncle ca. 6mm, deciduous; corolla spreading, white; corona red, slightly transparent.

Habits: Plant likes half-shading, warm and moist environment; leaf blade grow poorly in strong sun light; cold-resistant, proper decrement in the temperature can help blooming; loose and ventilate substrate preferred; mature branches are better for cuttings; fast-growing.

Note: Named after the famous German botanist, Adolph Engler.

Lighting requirement:
Maintenance difficulty:

Hoya erythrina Rintz in Malayan Nat. J. **30**(3-4): 501. 1978.

珊瑚红球兰

分布： 马来西亚。

形态特征： 攀援藤本。茎直径约2mm。叶片肉质，较硬，阔披针形至椭圆形，长约12cm，宽约5cm，边缘常波状，先端渐尖，深绿色，光照较强则变为深红色，表面常具斑点，叶脉明显。伞形花序具花7~12朵；花直径约1.4cm，几无香气；花冠平展状，先端略向背面反卷，黄色至黄绿色，先端有时淡棕色；副花冠先端内弯，黄色。

栽培习性： 喜半阴至明亮的环境；喜潮湿，但亦较耐旱；成熟枝条扦插易成活，但新芽分化较慢，新芽萌发后则生长快；新叶在强光及较大温差下极易变成红色。

光照需求：

养护难度：

Distribution: Malaysia.

Morphology: Climbing vine. Stem ca. 2mm diameter. Leaf blade fleshy, rigid, lanceolate to ellipse, ca. 12cm long and ca. 5cm wide, margin undulant, apex acuminate, dark green, turn dark red in strong sun light, often with spots on the surface; vein distinct. Umbel 7-12 flowered; flower ca. 1.4cm diameter, nearly odourless; corolla spreading, apex slightly revolute, yellow to yellowish green, sometimes apex light brown; corona apex recurved, yellow.

Habits: Plant likes half shade to bright space and moist environment, drought-resistant also; mature branch cuttings have high survival rate, new buds grow slowly but grows fast; new leaf blade tend to turns red in strong light and temperature difference.

Lighting requirement:

Maintenance difficulty:

Hoya erythrostemma Kerr in Bull. Misc. Inform. Kew, **1939**: 460. 1939.

红副球兰

分布：缅甸、泰国（南部）、马来西亚和印度尼西亚。

形态特征：攀援或悬垂藤本。茎直径约 2mm。叶片肉质，披针形，长约 8cm，宽约 5cm，边缘全缘，先端渐尖，新叶在较强光照下常呈红色，后变为浅绿色至绿色。伞形花序具花 20~40 朵；花直径约 1cm，具淡香气；花冠向背面反折，颜色多样，白色、淡黄白色、粉红色、红色、紫红色或蓝紫色，表面密被茸毛；副花冠淡黄白色至暗红色。

栽培习性：喜半阴；喜温暖潮湿的环境；扦插枝条生根很慢；植株生长缓慢，需适当增加施肥以促进生长及开花。

光照需求：☀ ☀ ☀

养护难度：🌱 🌱

Distribution: Myanmar, Thailand (southern region), Malaysia and Indonesia.

Morphology: Climbing or hanging vine. Stem ca. 2mm diameter. Leaf blade fleshy, lanceolate, ca. 8cm long and ca. 5cm wide, margin entire, apex acuminate, young leaves red in strong sun light and become light green to green. Umbel 20-40 flowered; flower ca. 1cm diameter, light fragrance; corolla reflexed, various colors, white, light yellowish white, pink, red, fuchsia or bluish violet, densely tomentose; corona light yellowish to dark red.

Habits: Plant likes half-shading, warm and moist environment; cuttings form roots slowly; slowly-growing, needs proper fertilizers to help its growing and blooming.

Lighting requirement: ☀ ☀ ☀

Maintenance difficulty: 🌱 🌱

Hoya excavata Teijsm. & Binn. in Natuurk. Tijdschr. Ned.-Indië, 25: 406. 1863.

凹副球兰

分布：印度尼西亚（加里曼丹岛和苏拉威西岛）。

形态特征：攀援藤本。茎直径约4mm。叶片肉质，厚，椭圆形，长约14cm，宽约8cm，边缘全缘，先端圆钝，深绿色，叶脉不明显。伞形花序呈球状，具花10~25朵；花直径约1.5cm，无香气；花冠平展状，边缘略向背面反卷，粉红色、红色、深红色至紫红色，表面具短柔毛；副花冠红色至紫红色。

栽培习性：植株健壮，适应性强；喜半阴；可耐一定的干旱和低温；扦插枝条易生根，但须待植株生长健壮后才会开花。

光照需求：

养护难度：

Distribution: Indonesia (Kalimantan Island and Sulawesi Island).

Morphology: Climbing vine. Stem ca. 4mm diameter. Leaf blade fleshy, thick, ellipse, ca. 14cm long and ca. 8cm wide, margin entire, apex obtuse, deep green, vein obscure. Umbel convex, 10-25 flowered; flower ca. 1.5cm diameter, odourless; corolla spreading, corolla margin slightly revolute, pink, red, deep red to fuchsia, tomentose; corona red to fuchsia.

Habits: Healthy and well-adapted; half shade environment preferred; endure certain drought and low temperature; cuttings form roots easily; plant needs a long period to blossom after plant grows maturation.

Lighting requirement:

Maintenance difficulty:

Hoya finlaysonii Wight, Contr. Bot. India, 38. 1834.

芬莱森球兰 斐赖迅球兰

分布：泰国、马来西亚和印度尼西亚。

形态特征：攀援藤本。茎直径约 3mm。叶片肉质，椭圆形至椭圆状披针形，长约 15cm，宽约 5cm，边缘波状或具圆齿，先端渐尖至尾尖，下面微红，上面浅绿色，有时具银色斑点，叶脉深绿色，通常明显，稀不明显。伞形花序呈球状，具花 15~40 朵；花直径约 8mm，具淡柑橘香气；花冠向背面反卷，淡黄色，先端常红褐色；副花冠白色。

栽培习性：适应性强；喜光；喜温暖潮湿的环境，亦较耐旱；植株生长较缓慢，光照充足可促进生长和开花。

备注：本种以苏格兰博物学家 George Finlayson 命名。

光照需求：

养护难度：

Distribution: Thailand, Malaysia and Indonesia.

Morphology: Climbing vine. Stem ca. 3mm diameter. Leaf blade fleshy, ellipse to ellipse lanceolate, ca. 15cm long and ca. 5cm wide, margin undulant or crenate, apex acuminate to caudate, abaxially light red, adaxially light green and sometimes with silver spots, vein deep green, generally instinct, rare obscure. Umbel convex, 15-40 flowered; flower ca. 8mm diameter, citrus fragrance; corolla revolute, light yellow, corolla apex often red brown; corona white.

Habits: Well-adapted and likes sun light; warm and moist environment preferred; drought-resistant; slow-growing; sufficient sun light can help its growing and blooming.

Note: Named after the Scottish naturalist George Finlayson.

Lighting requirement:

Maintenance difficulty:

Hoya fitchii Kloppenb. in Fraterna 22(4): 16. 2009.

费氏球兰

分布：菲律宾（吕宋岛）、马来西亚和印度尼西亚。

形态特征：攀援藤本，茎直径约3mm。叶片薄草质，长卵形至椭圆形，长约5cm，宽约3cm，边缘全缘，先端渐尖，深绿色，叶脉明显。伞形花序具花15~25朵；花直径约1cm，具玫瑰香气；花冠向背面反折，淡黄褐色，边缘褐色；副花冠淡黄白色；雄蕊红色。

栽培习性：适应性强，易养护，喜半阴，较耐寒；扦插枝条不易生根，但一旦根系生长稳定则植株生长快；适当施肥可促进植株开花。

备注：本种以园艺学家Charles M. Fitch命名。

光照需求：

养护难度：

Distribution: Philippines (Luzon Island), Malaysia and Indonesia.

Morphology: Climbing vine. Stem ca. 3mm diameter. Leaf bladethin leathery, oblong to ellipse, ca. 5cm long and ca. 3cm wide, margin entire, apex acuminate, deep green, vein distinct. Umbel 15-25 flowered; flower ca. 1cm diameter, rose fragrance; corolla reflexed, light yellowish brown, corolla margin brown; corona light yellowish white; stamen red.

Habits: Well-adapted and easy-cultivated; likes half shade; cold-resistant; cuttings form rooting rarely, but grows fast after forming roots; proper fertilizers can help its blooming.

Note: Named after the horticulturalist, Charles M. Fitch.

Lighting requirement:

Maintenance difficulty:

Hoya flagellata Kerr in Hooker's Icon. Pl. **35**: t. 3407. 1950.

鞭花球兰

分布：泰国和马来西亚。

形态特征：攀援藤本。茎直径约2mm。叶片肉质，较硬，披针形，长约4~6cm，宽约2~3cm，边缘全缘，先端渐尖，下面常红色，上面绿色并具银色斑点，表面被短柔毛。伞形花序半球形，具花6~12朵；花直径约1.2cm，几无香气；花冠平展，先端长尾状并向背面反卷，白色或浅粉色，背面被绢毛；副花冠红色；雄蕊先端具5条毛状附属物。

栽培习性：喜光；较强光照下叶片红色明显；喜温暖潮湿的环境；不耐寒，气温8℃以下易死亡；栽培宜用疏松、透气的基质；扦插枝条易成活，但生长较缓慢；植株需较长的生长期才能开花。

光照需求：
养护难度：

Distribution: Thailand and Malaysia.

Morphology: Climbing vine. Stem ca. 2mm diameter. Leaf blade fleshy, stiff, lanceolate, 4-6cm long and 2-3cm wide, margin entire, apex acuminate, abaxially red, adaxially green, with silver spots, tomentose. Umbel convex, 6-12 flowered; flower ca. 1.2cm diameter, nearly odourless; corolla spreading, apex caudate and revolute, white or light pink, abaxially sericeous; corona red; stamen apex has hairy appendage.

Habits: Plant likes sun light, warm and moist environment; leaf blade red in strong sun light; nonhardy, and tends to die when temperature drops below 8 C ; loose and ventilate substrate preferred; cuttings have high survival rate, grow slowly; needs a long growing period to give blossom.

Lighting requirement:
Maintenance difficulty:

Hoya flavida P. I. Forst. & Liddle in Austrobaileya, 4(1): 53–55. 1993.

淡黄球兰

分布：所罗门群岛。

形态特征：攀援藤本。茎直径约 2mm。叶片形状变化较大，通常为椭圆体形至阔披针形，长约 10cm，宽约 3~4cm，幼叶常为红色，成熟后常为浅绿色至深绿色，如阳光强烈则为红色，中脉在上面突起，侧脉不显。伞形花序具花 10~15 朵；花直径约 1.5cm，具淡香气；花冠平展，先端略向背面反卷，橙色，边缘具白色茸毛；副花冠黄色；雄蕊红色。

栽培习性：不耐寒，冬季需保温，10℃以下即受寒害并干枯；栽培宜用疏松、透气的基质；扦插宜用成熟枝条，扦插枝条易成活；植株较难开花。

光照需求：

养护难度：

Distribution: Solomon Islands.

Morphology: Climbing vine. Stem ca. 2mm diameter. Leaf blade various shaped, ellipse to wide lanceolate, ca. 10cm long and 3-4cm wide, young leaves red in color, become light green to dark green in mature stage, or keep red in strong sun light; adaxially midvein protuberant, lateral vein obscure. Umbel 10-15 flowered; flower ca. 1.5cm diameter, light fragrance; corolla spreading, corolla apex revolute, orange, margin white tomentose; corona yellow; stamen red.

Habits: Nonhardy, need insulation in cold seasons; tends to suffer from chilling injury and wither when temperature drops below 10 C ; loose and ventilate substrate preferred; mature branches as cuttings have high survival rate; difficult to bloom.

Lighting requirement:

Maintenance difficulty:

Hoya fraterna Blume, Rumphia, **4**: 32. 1849.

香水球兰

分布：印度尼西亚（爪哇岛）。

形态特征：攀援藤本，茎直径约 3mm。叶片薄，椭圆体形，长约 11cm，宽约 6cm，深绿色，叶脉明显。伞形花序大，具花 30~40 朵；花直径 1.7cm，具香气；花冠向背面反折，杏黄色至金黄色，被茸毛；副花冠米黄色。

栽培习性：喜半阴；喜温暖潮湿的环境，但亦较耐干旱；栽培宜选择疏松透气的基质；成熟枝条扦插易生根；初期生长较缓慢，植株健壮后则生长较快，花序萌生早，容易开花。

光照需求：

养护难度：

Distribution: Indonesia (Java Island).

Morphology: Climbing vine. Stem ca. 3mm diameter. Leaf blade thin, ellipse, ca. 11cm long and ca. 6cm wide, deep green, vein distinct. Umbel large, 30-40 flowered; flower ca. 1.7cm diameter, fragrance; corolla reflexed, apricot to golden yellow, tomentose; corona beige.

Habits: Grows well in half-shading, warm and moist environment; drought-resistant; loose and ventilate substrate preferred; mature branch cuttings form roots easily; slowly-growings at early stage, and fast-growing when mature; bores inflorescence early, and blooms easily.

Lighting requirement:

Maintenance difficulty:

Hoya fungii Merr. in Lingnan Sci. J. **13**(1): 68–69. 1934.

护耳草

分布：中国（广东、广西、海南和云南）。

形态特征：攀援藤本。茎直径约 3.5mm。叶片肉质，椭圆形，长约 11cm，宽约 7cm，边缘全缘，先端具小尖头，绿色，叶脉明显。伞形花序大，呈球状，具花 25~35 朵；花直径约 1.8cm，具浓烈香气，夜晚更甚；花冠平展，白色，被短柔毛；副花冠淡黄白色，雄蕊红色。

栽培习性：适应性强；喜半阴；耐寒冷；对栽培基质要求不高，仅需透气即可；扦插枝条易成活；新萌发的枝条长至 1 m 左右即可开花，花期长且一年反复开花 3~5 次。

备注：本种以我国植物采集人冯钦（Fung H.）命名。

光照需求：

养护难度：

Distribution: China (Guangdong, Guangxi, Hainan and Yunnan).

Morphology: Climbing vine. Stem ca. 3.5mm diameter. Leaf blade fleshy, ellipse, ca. 11cm long and ca. 7cm wide, margin entire, apex mucronulate, green, vein distinct. Umbel large, convex, 25-35 flowered; flower ca. 1.8cm diameter, strong fragrance, and stronger at night; corolla spreading, white, puberulent; corona light yellowish white; stamen red.

Habits: Well-adapted and likes half shade; cold-resistant; low demand in substrate; ventilate substrate preferred; cuttings have high survival rate; new-born branches can bloom when reaches to 1m long; longer blossom period is 3 to 5 times a year.

Note: Named after the Chinese collector, Fung H..

Lighting requirement:

Maintenance difficulty:

Hoya fusca Wall.Pl. Asiat.Rar. **1**: 68, pl. 75. 1830.

黄色球兰（新拟） 黄花球兰

分布：中国、印度（北部）、不丹、尼泊尔、缅甸、老挝、柬埔寨、泰国和越南。

形态特征：半灌木。茎直径约 2.5cm，老茎常木质化。叶片革质，长椭圆形至披针形，长 10~13cm，宽 2.5~4.5cm，先端具小尖头，绿色，中脉在叶片上面下凹，侧脉垂直于主脉。伞形花序具花 9~16 朵；花直径 1~1.5cm；花冠深裂至基部，常向背面反折，亮黄色；副花冠黄色至粉红色。

栽培习性：喜半阴；较耐寒，可耐受 5℃左右的低温；栽培需用透气、潮湿但不积水的基质；扦插宜用木质化的老茎；如环境适合则生长较快，但需较长的营养生长期才可开花，且植株花序较少。

光照需求：
养护难度：

Distribution: China, India (northern region), Bhutan, Nepal, Myanmar, Laos, Combodia, Thailand and Vietnam.

Morphology: Subshrub. Stem ca. 2.5cm diameter; mature stems often lignified. Leaf blade leathery, oblong to lanceolate, 10-13cm long and 2.5-4.5cm wide, apex mucronulate, green, midvein concave; lateral veins perpendicular to midvein. Umbel 9-16 flowered; flower 1-1.5cm diameter; corolla split to the base, often reflexed, bright yellow; corona yellow to pink.

Habits: Plant likes half shade, cold-resistant and can endure a low temperature at 5℃; ventilate, moist and non-water-logging substrate preferred; slight lignified stems as cuttings; fast-growing in suitable environment, needs a long vegetative period to blossom; inflorescence not plenty as other species.

Lighting requirement:
Maintenance difficulty:

Hoya glabra Schltr. in Bot. Jahrb. Syst. 40(Beibl. 92): 14. 1908.

光叶球兰

分布： 马来西亚。

形态特征： 攀援藤本。茎直径约 4mm。叶片肉质，阔卵形，长约 12cm，宽约 8cm，先端渐尖，淡绿色至绿色，有时具银色斑点，叶脉在两面突起，明显。伞形花序呈球状，具花 26~35 朵；花直径约 1cm，具淡香气；花冠平展，先端常向背面反卷，紫红色至红褐色，先端常粉红色；副花冠粉红色。

栽培习性： 喜明亮的散射光；不耐低温，环境温度低于 10℃需保温；扦插宜采用成熟枝条，较易成活，但叶片生长缓慢，生长期需追肥以促进叶片生长；本种扦插的枝条常先长出极长的枝条后才开始长叶；健壮植株一年可多次开花；由于植株较大，需预留较大空间供其生长。

光照需求：

养护难度：

Distribution: Malaysia.

Morphology: Climbing vine. Stem ca. 4mm diameter. Leaf blade fleshy, wide ovate, ca. 12cm long and ca. 8cm wide, apex acuminate; light green to green, sometimes with silver spots on the surface; vein distinct and protuberant on both sides. Umbel convex, 26-35 flowered; flower ca. 1cm diameter, light fragrance; corolla spreading, apex often revolute, fuchsia to red brown, apex often pink; corona pink.

Habits: Plant likes bright scattered light; nonhardy, needs insulation when temperature drops to below 10℃; mature branches as cuttings have high survival rate; leaves grow slowly, to help the leaf's growing, additional fertilizers needed during growing period; cuttings often bore long branches, then leaves grow; healthy plants can bloom several times per year; due to the large plant shape, enough growing space preferred.

Lighting requirement:

Maintenance difficulty:

Hoya globulifera Blume, Mus. Bot. **1**: 44. 1850.

球芯球兰

分布：巴布亚新几内亚。

形态特征：攀援藤本。茎直径约 2mm，节间较长。叶片略肉质，较硬，椭圆形至披针形，长约 10cm，宽约 3cm，边缘全缘，先端渐尖，新叶常呈浅红色，老叶深绿色。伞形花序具花 20~30 朵；花直径约 1.5cm，几无香气；花冠平展，深红褐色，表面被短茸毛；副花冠深红色。

栽培习性：喜阴，适当遮阴有利于生长；喜潮湿且通风的环境，可延长花期；成熟枝条扦插易成活；植株生长较快，需经常整形以保持株形。

光照需求：

养护难度：

Distribution: Papua New Guinea.

Morphology: Climbing vine. Stem ca. 2mm diameter, internode long. Leaf blade fleshy, rigid, ellipse and lanceolate, ca. 10cm long and ca. 3cm wide, margin entire, apex acuminate, tender leaves light red and mature ones deep green. Umbel 20-30 flowered; flower ca. 1.5cm diameter, nearly odourless; corolla spreading, deep red brown, tomentose; corona deep red.

Habits: Plant likes shad to grow; moist and ventilate environment can prolong the florescence; mature branch cuttings have high survival rate; fast-growing; frequent trimming is needed to maintain the plant shape.

Lighting requirement:

Maintenance difficulty:

Hoya globulosa J. D. Hook. in Gard. Chron. **1**: 732. 1882.

球序球兰（新拟）小球球兰

分布：印度、缅甸和泰国。

形态特征：攀援藤本。茎直径约 3.5mm，被短柔毛。叶片略肉质，常为长圆形，长约 15cm，宽约 5cm，光照不足时叶片较长且较窄，边缘波状，先端具小尖头，浅绿色，表面被短柔毛，叶脉深绿色，明显。伞形花序呈球状，具花 25~35 朵；花直径约 1.2cm，具香气；花冠平展或略向背面反卷，黄白色；副花冠白色，边缘常红色。

栽培习性：适应性强；对光照要求不高，适当增加光照有利于生长及花序萌生；耐寒冷，可耐受 5℃的低温；扦插枝条易生根，生长较快；本种开花较少，但叶片美观，可作为观叶种类栽培。

光照需求：
养护难度：

Distribution: India, Myanmar and Thailand.

Morphology: Climbing vine. Stem ca. 3.5mm diameter, puberulent. Leaf blade fleshy, ovate, ca. 15cm long and ca. 5cm wide, or longer and narrower in insufficient of sun light; margin undulant, apex mucronulate, light green, puberulent, vein deep green, distinct. Umbel convex, 25-35 flowered; flower ca. 1.2cm diameter, fragrance; corolla spreading, revolute, yellowish white; corona white, margin red.

Habits: Well-adapted and low demand in sun light; proper increment in sun light help its growing and to form inflorescence; cold-resistant, endure a low temperature below to 5℃; cuttings are easily form roots and growing fast; blooms fewer flowers than other species, but leaves beautiful and have great ornamental value.

Lighting requirement:
Maintenance difficulty:

Hoya golamcoana Kloppenb.in Fraterna, 1991(3), Philipp. Hoya Sp. Suppl.: II. 1991.

格兰柯球兰

分布：菲律宾（巴拉望群岛）。

形态特征：攀援藤本。茎直径约 3mm，节间较长，老茎常呈木质化。叶片薄草质，较厚实，椭圆形，长约 4cm，宽约 2.5cm，深绿色，中脉浅绿色，较明显。伞形花序具花 10~15 朵，开花后花序梗常同花一起脱落；花直径约 1.2cm，具甜香气；花冠向背面反折，白色至浅绿色；副花冠基部常粉红色，边缘白色。

栽培习性：适应性强；喜光照充足的环境，如置于荫蔽环境则节间伸长；扦插枝条易成活，生长快；需经过较长的营养生长期才可开花；老枝的叶片常脱落，可将枝条相互缠绕增加植株观赏性。

备注：本种以菲律宾园艺学家 Jun Golamco 命名。

光照需求：☀☀☀☀☀

养护难度：🌱

Distribution: Philippines (Palawan Island).

Morphology: Climbing vine. Stem ca. 4mm diameter, internode long; mature stems often lignified. Leaf blade thin, leathery, thick, ellipse, ca. 4cm long and ca. 2.5cm wide, deep green, midvein light green, distinct. Umbel 10-15 flowered; flower ca. 1.2cm diameter, sweet fragrance; peduncle drops after flowers wither; corolla reflexed, white to light green; corona base often pink, margin white.

Habits: Well-adapted and grows well in sufficient sun light; internode grows longer in shade space; cuttings have high survival rate and are growing fast; needs a long vegetative period to blossom; mature leaves often fall off; in order to increase ornamental value, branches can be intertwined.

Note: Named after the Philippines horticulturalist, Jun Golamco.

Lighting requirement: ☀☀☀☀☀

Maintenance difficulty: 🌱

Hoya graveolens Kerr in Bull. Misc. Inform. Kew, **1939**: 461. 1939.

烈味球兰

分布：泰国。

形态特征：攀援藤本。茎直径约2.5mm，较硬，老茎木质化。叶片肉质，深绿色，表面偶见银色斑点，有倒卵形和披针形两种叶型，倒卵形的叶片长约10cm，宽约6cm，先端急尖，披针形的叶片长约11cm，宽约3cm，先端渐尖。伞形花序具花13~20朵；花直径约1.5cm，具浓香气；花冠平展，先端常内弯，白色，表面被亮绢毛；副花冠白色至粉红色；雄蕊红色。

栽培习性：植株适应性较强，但须避免直射光及干燥的环境；成熟枝条扦插易生根；成活后生长快速。

光照需求：

养护难度：

Distribution: Thailand.

Morphology: Climbing vine. Stem ca. 2.5mm diameter; mature stems lignified. Leaf blade fleshy, deep green, occasionally with silver spots, with two kinds of shape: obovate one ca. 10cm long and ca. 6cm wide, apex acute; lanceolate one ca. 11cm long and ca. 3cm wide, apex acuminate. Umbel 13-20 flowered; flower ca. 1.5cm diameter, strong fragrance; corolla spreading, apex often recurved, white, sericeous; corona white to pink; stamen red.

Habits: Well-adapted; avoid direct sun light and dry environment; mature branch cuttings form roots easily, and grow fast.

Lighting requirement:

Maintenance difficulty:

Hoya greenii Kloppenb.in Fraterna, **1995**(2): 12. 1995.

格林球兰

分布：菲律宾（棉兰老岛）。

形态特征：攀援藤本，枝条常披散。茎直径约 3mm。叶片薄，披针形，长约 13cm，宽约 3cm，边缘全缘，先端渐尖，绿色，表面粗糙，中脉明显。伞形花序具花 25~35 朵；花直径约 1.8cm，几无香气；花冠向背面反折，淡粉色；副花冠粉色，雄蕊红色。

栽培习性：喜温暖潮湿的环境，不耐干旱；扦插枝条置于高温潮湿的环境中成活率高，且生长快；需经过较长的营养生长期才可开花；花期较长。

备注：本种以美国园艺学家及采集人 Ted Green 命名。

光照需求：☀☀☀

养护难度：🌱🌱🌱

Distribution: Philippines (Mindanao Island).

Morphology: Climbing vine, with sprawling branches. Stem ca. 3mm diameter. Leaf blade thin, lanceolate, ca. 13cm long and ca. 3cm wide, margin entire, apex acuminate, green, surface rough, midvein distinct. Umbel 25-35 flowered; flower ca. 1.8cm diameter, nearly odourless; corolla reflexed, light pink; corona pink; stamen red.

Habits: Plant prefers warm and moist environment; drought-sensitive; cuttings have high survival rate in warm and moist environment; grow very fast; needs long vegetative period to blossom; long blossom period.

Note: Named after the American horticulturalist and collector Ted Green.

Lighting requirement: ☀☀☀

Maintenance difficulty: 🌱🌱🌱

Hoya griffithii J. D. Hook. Fl. Brit. India, **4**(10): 59. 1883.

荷秋藤

分布：中国（南部和西南部）和印度。

形态特征：攀援藤本。茎直径约4cm。叶片肉质，披针形至椭圆形，长11~14cm，宽2.5~4.5cm，先端渐尖至急尖，深绿色，中脉明显。伞形花序具花8~12朵；花直径约3cm，具浓香气；花冠常向内略呈爪状，初期浅粉色，后变为黄色，外面常具紫红色斑点；副花冠白色直至粉色。

栽培习性：栽培宜置于明亮的散射光下；在潮湿但不积水的细小基质中生长较佳；采用成熟的枝条扦插较易生根，初期需保护根部避免受损，否则扦插枝条易干枯；根部充分发育后植株生长快；健壮植株开花频繁。

备注：本种以英国植物学家和医生William Griffith命名。

光照需求：

养护难度：

Distribution: China (southern and southwest region) and India.

Morphology: Climbing vine. Stem ca. 4mm diameter. Leaf blade fleshy, lanceolate to ellipse, 11-14cm long and 2.5-4.5cm wide, apex acuminate to acute, deep green, midvein distinct. Umbel 8-12 flowered; flowerca. 3cm diameter, strong fragrance; corolla incurved, claw-like, light pink at early stage, later turns into yellow, often fuchsia spots on the surface; corona white to pink.

Habits: Plant likes bright scattered light; moist but non-water-logging substrate preferred; mature branch cuttings raise roots easily; avoid root damages at early stage, otherwise cuttings tend to wither; fast-growing after roots formation; healthy plant blooms frequently.

Note: Named after the British botanist and doctor, William Griffith.

Lighting requirement:

Maintenance difficulty:

Hoya halconensis Kloppenb.in Fraterna **1**(3), Philipp. Hoya Sp. Suppl.: III. 1990.

阿尔孔球兰

分布：菲律宾（民都洛岛）。

形态特征：攀援藤本。茎直径约5mm。叶片厚纸质至革质，披针形，长约12cm，宽约5cm，边缘全缘，先端渐尖，绿色，叶脉较明显。伞形花序呈球状，具花15~25朵；花直径约2cm，具奶油香气；花冠平展或略内弯，浅黄色至黄色，被短茸毛；副花冠中间红色，边缘黄绿色。

栽培习性：喜半阴；喜高温高湿的环境，湿度不足则叶片容易干枯；栽培宜选用疏松、透气且保湿的基质；扦插枝条易生根，植株生长较快，需要多施肥支持其生长；本种花期较短，花开放3~4天后即脱落。

备注：本种以其产地菲律宾阿尔孔山（Mount Halcon）命名。

光照需求：

养护难度：

Distribution: Philippines (Mindoro Island).

Morphology: Climbing vine. Stem ca. 5mm diameter. Leaf blade thick papyraceous to leathery, lanceolate, ca. 12cm long and ca. 5cm wide, margin entire, apex acuminate, green, vein distinct. Umbel convex, 15-25 flowered; flower ca. 2cm diameter, vanilla fragrance; corolla spreading or slightly recurved, light yellow to yellow, tomentose; corona center red, margin yellowish green.

Habits: Grows well in half shade and environment of high temperature and high humidity; leaf blades tend to wither in low hummidity; loose, ventilate and moist substrate preferred; cuttings form roots easily; fast-growing and needs more fertilizer to promote its growth; florescence short; flowers wither after 3-4 days.

Note: This *Hoya* comes from Mount Halcon, Philippine and named from this place.

Lighting requirement:

Maintenance difficulty:

Hoya heuschkeliana Kloppenb. in Hoyan, **11**(1:2): i. 1989.

休斯科尔球兰

分布：菲律宾。

形态特征：攀援或垂吊藤本。茎直径约 2mm。叶片稍肉质，小，卵圆形，长约 2cm，宽约 1.5cm，先端急尖，下面浅绿色，上面深绿色，叶脉不明显。伞形花序具花 5~9 朵；花直径约 5mm，气味酸；花冠呈坛状，先端常外卷，粉红色或橙黄色；副花冠粉红色或黄色。

栽培习性：喜半阴至明亮的环境，在明亮环境下叶片较小且厚实，半阴环境下则较大且薄；喜温暖潮湿的环境；扦插枝条易生根；生长快，缠绕或垂吊栽培均可；开花早，如环境适宜则终年开花不断；本种花气味较独特，如不适宜则避免在室内栽培。

备注：本种以植物采集人 Dexter Heuschkel 命名，具有两种不同花色，常被分为两个栽培品种：粉花休斯科尔球兰 *Hoya heuschkeliana* **'Pink'** 和黄花休斯科尔球兰 *Hoya heuschkeliana* **'Yellow'**，其主要区别是前者叶片较柔软，边缘平展，花冠粉红色；后者叶片较硬，边缘向下反卷，花冠黄色。

光照需求：☀☀☀

养护难度：🌱🌱

Distribution: Philippines.

Morphology: Climbing or hanging vine. Stem ca. 2mm diameter. Leaf blade fleshy, small, ovate, ca. 2cm long and ca. 1.5cm wide, apex acute, abaxially light green, adxially deep green, vein obscure. Umbel 5-9 flowered; flower ca. 5mm diameter, with sourodour; corolla urceolate, apex revolute, pink to orange; corona pink or yellow.

Habits: Plant likes half shade to bright environment; leaf blade thick and small in bright environment, thin and large in half shade; warm and moist environment preferred; cuttings fromroots easily; fast-growing; twining or hanging cultivation pattern preferred; blooms early; bloom throughout the year in suitable environment; due to special odour, avoid to grow at indoors if people uncomfortable.

Note: Named after the plant collector Dexter Heuschkel. *Hoya heuschkeliana* is classified into two cultivarsbase on its two different flower colors: 'Pink' and 'Yellow'. Former one has soft leaves with explanate margin and pink corolla; later one has rigid leaves with warped margin and yellow corolla.

Lighting requirement: ☀☀☀

Maintenance difficulty: 🌱🌱

Hoya hypolasia Schltr. in Bot. Jahrb. Syst. **50**: 123. 1913.

绒叶球兰

分布：印度尼西亚和巴布亚新几内亚。

形态特征：攀援藤本，枝条常外展。茎直径约2mm。叶片稍肉质，较硬，披针形，长约12cm，宽约2.5cm，深绿色，被短茸毛，叶脉浅绿色，中脉在上面下凹，侧脉明显。伞形花序具花15~25朵；花直径约2cm，无香气；花冠平展或略内弯，浅黄色，被微柔毛；副花冠黄色。

栽培习性：昼夜温差较大可促进植株开花；喜潮湿且通风的环境；栽培基质需保持适度潮湿，过于湿润或干燥均不利于生长；扦插宜选用略成熟的枝条，老枝或嫩枝扦插均不易生根；环境适合且根系健壮则生长极快。

光照需求：
养护难度：

Distribution: Indonesia and Papua New Guinea.

Morphology: Climbing vine, with spreading branches. Stem ca. 2mm diameter. Leaf blade fleshy, rigid, lanceolate, ca. 12cm long and ca. 2.5cm wide, deep green, tomentose, vein light green, midvein concave, lateral vein distinct. Umbel 15-25 flowered; flower ca. 2cm diameter, odourless; corolla spreading or slightly incurved, light yellow, puberulent; corona yellow.

Habits: Plant likes ventilate and moist environment; keep proper humidity of the substrate; too wet or too dry environmentwill impede to grow; mature branch better for cuttings, old or young branches rarely form roots; grow fast in suitable environment; temperature fluctuation between day and night can promote it blooming.

Lighting requirement:
Maintenance difficulty:

Hoya imbricata Callery ex Decne. in Prodr. 8: 637. 1844.

覆叶球兰 龟壳叶球兰

分布： 菲律宾、马来西亚和印度尼西亚（苏拉威西岛）。

形态特征： 攀援藤本。茎直径约 2mm。两枚对生叶中的一枚常不发育；叶片薄革质，圆形，直径约 7cm，深绿色，叶脉不显。伞形花序具花 13~25 朵；花直径约 1cm，具淡香气；花冠中部以下平展，先端向背面反卷，有时全部向背面反卷，花冠浅绿色至浅黄色，被短茸毛。副花冠浅黄绿色，有时间杂淡粉色；雄蕊上具 5 条丝状毛。

栽培习性： 喜高湿环境；扦插枝条易生根；本种叶片紧贴附生物表面，如无附生物贴附则闭合呈饺子状，新芽、花序及不定根均从叶片覆盖的节上伸出；需经过较长的营养生长期，待植株健壮后才会开花。

光照需求：

养护难度：

Distribution: Philippines, Malaysia and Indonesia (Sulawesi Island).

Morphology: Climbing vine. Stem ca. 2mm diameter. Only one leaf develop per nod; leaf blade thin leathery, round, ca. 7cm diameter, deep green, vein obscure. Umbel 13-25 flowered; flower ca. 1cm diameter, light fragrance; corolla spreading below the middle, apex revolute, sometimes entirely revolute, light green to light yellow, tomentose; corona light yellowish green, sometimes pink-mixed; five silky hairs on the top of each stamen.

Habits: Plant likes environment of high humidity; cuttings form roots easily; leaves tightly cling to the surface of supports, otherwise close like dumpling; young buds, inflorescence and adventive roots extend from the nod covered by the leaves; due to the long vegetative period, needs certain period to bloom.

Lighting requirement:

Maintenance difficulty:

Hoya imbricata f. *basi-subcordata* Koord. in Philipp. J. Sci. **15**: 264. 1919.

基心覆叶球兰 玳瑁叶球兰

分布：马来西亚和菲律宾。

形态特征：本变形与原种的区别在于本变形叶片表面满布银色斑点。

栽培习性：同原种。

光照需求：●●●

养护难度：🌱🌱🌱

Distribution: Malaysia and Philippines.

Morphology: Difference between original specie and this forma is silver spots all-over the leaf surface.

Habits: Same as original species.

Lighting requirement: ●●●

Maintenance difficulty: 🌱🌱🌱

Hoya imbricata f. *basi-subcordata* 'Red'

红花基心覆叶球兰

形态特征：本栽培品种与基心覆叶球兰的区别在于叶片较大，直径约 8cm，表面银色斑点更大，呈斑块状；副花冠红色至粉红色。

栽培习性：同原种。

光照需求：

养护难度：

Morphology: Larger leaf, ca. 8cm; silver spots on leaf surface are larger, plaque-like; corona red to pink.

Habits: Same as original species.

Lighting requirement:

Maintenance difficulty:

Hoya imperialis Lindl.in Edwards's Bot. Reg. **33**: sub t. 68. 1847.

帝王球兰 红花帝王球兰

分布：菲律宾和马来西亚。

形态特征：攀援藤本。茎肉质，直径约5mm。叶片薄，较硬，到卵状披针形、披针形至椭圆形，长10~15cm，宽4~5cm，边缘波状，先端急尖至渐尖，深绿色，中脉明显。伞形花序具花5~12朵；花直径约6~8cm，无香气；花冠内弯呈爪状，紫红色，近副花冠处被短柔毛；副花冠淡黄白色。

栽培习性：喜光，充足的光照能促进开花；喜温暖潮湿的环境；不耐寒，12℃以下即受寒害，冬季忌吹风；扦插宜选用成熟枝条，新根常从创口长出，初期生长较慢。

光照需求：☀☀☀☀☀

养护难度：🌱🌱🌱🌱

Distribution: Philippines and Malaysia.

Morphology: Climbing vine. Stem ca. 4mm diameter, fleshy. Leaf blade thin, rigid, ovate to lanceolate, 10-15cm long and 4-5cm wide, margin undulant, apex acute to acuminate, deep green, midvein distinct. Umbel 5-12 flowered; flower 6-8cm diameter, odourless; corolla incurved, claw-like, fuchsia, inner base puberulent; corona light yellowish white.

Habits: Plant likes sun light; sufficient sun light help for blooming; warm and moist environment preferred; nonhardy, tends to suffer from chilling injury when temperature drops to below 12℃; avoid draught in winter; mature branches are better for cuttings; new roots usually grow from the cut place; cuttings grow slowly at early stage.

Lighting requirement: ☀☀☀☀☀

Maintenance difficulty: 🌱🌱🌱🌱

Hoya imperialis 'Pink'

粉花帝王球兰

形态特征：本栽培种与原种的区别主要在于花冠为淡紫红色。

栽培习性：同原种。

光照需求：☀ ☀ ☀ ☀

养护难度：🌱 🌱 🌱 🌱

Morphology: Corolla of this cultivar is light fuchsia, which is different from original species.

Habits: Same as original species.

Lighting requirement: ☀ ☀ ☀ ☀

Maintenance difficulty: 🌱 🌱 🌱 🌱

Hoya imperialis 'White'

白花帝王球兰

形态特征：本栽培种与原种的区别主要在于花冠为白色至淡黄白色；另外本种的花具香气，夜晚更甚。

栽培习性：同原种。

光照需求：

养护难度：

Morphology: Corolla of this cultivar is white to light yellowish white; moreover, this cultivar's flower has fragrance and especially stronger at night.

Habits: Same as original species.

Lighting requirement:

Maintenance difficulty:

Hoya incrassata Warb.in Perkkins, Fragm. Fl. Philipp. 1: 130.1904.

厚冠球兰

分布：泰国、菲律宾和马来西亚。

形态特征：攀援藤本，茎直径约 3cm。叶片薄革质，柔软，椭圆形至椭圆状披针形，长约 11cm，宽约 5cm，边缘全缘，先端急尖，淡绿色，中脉明显。伞形花序呈球状，具花 25~40 朵；花直径约 1cm，具浓烈的柑橘香气；花冠向背面反折或反卷，黄色，边缘紫红色至褐色；副花冠白色，半透明状。

栽培习性：适应性强；喜半阴；喜温暖潮湿的环境；栽培宜选择疏松、透气的基质；扦插选用成熟枝条成活率高，初期生长较慢；后期适当施肥则生长极快。

光照需求：

养护难度：

Distribution: Thailand, Philippines and Malaysia.

Morphology: Climbing vine. Stem ca. 3cm diameter. Leaf blade thin leathery, soft, ellipse to ellipse lanceolate, ca. 11cm long and ca. 5cm wide, margin entire, apex acute, light green, midvein distinct. Umbel convex, 25-40 flowered; flower ca. 1cm diameter, strongcitrus fragrance; corolla reflexed or revolute, yellow, margin fuchsia to brown; corona white, translucent.

Habits: Well-adapted and grows well in half-shading, warm and moist environment; loose and ventilate substrate preferred; mature branch cuttings have high survival rate, but grow slowly at early stage; after early stage grows very fast when proper fertilizers given.

Lighting requirement:

Maintenance difficulty:

Hoya incrassata 'Eclipse'

厚冠球兰"日蚀"

形态特征： 本栽培种与原种的区别在于叶片较硬，边缘白色并向背面反卷。

栽培习性： 同原种。

光照需求： ●●●

养护难度： 🌱

Morphology: Leaf of this cultivar is harder; leaf margin is white and revolute.

Habits: Same as original species.

Lighting requirement: ●●●

Maintenance difficulty: 🌱

Hoya incrassata 'Moon Shadow'

厚冠球兰"月影"

形态特征：本栽培种与原种的区别在于叶片中间具白色斑块。

栽培习性：同原种。

光照需求：☀☀☀

养护难度：🍴

Morphology: This cultivar has white spots on leaf center, which is different from original species.

Habits: Same the original species.

Lighting requirement: ☀☀☀

Maintenance difficulty: 🍴

Hoya javanica Boerl.in Handl. Fl. Ned. Ind. 2: 440. 1899.

爪哇球兰

分布：马来西亚和印度尼西亚（爪哇岛）。

形态特征：直立半灌木。茎木质化，直径约4cm。叶片纸质，薄，长椭圆形，长约11cm，宽约5cm，边缘波状，先端急尖，绿色，有时具银色斑点，中脉明显，侧脉不显。伞形花序具花20~40朵；花直径约1.6cm，具淡柠檬香气；花冠向背面反折，中间白色，分裂部分浅黄色至黄色；副花冠白色，突出，基部具短茸毛；雄蕊黑色。

栽培习性：适当加强光照能使叶片生长更佳及促进开花；不耐干旱，需勤浇水；栽培宜选用疏松的基质；取成熟枝条扦插较易生根。

备注：本种以其产地印度尼西亚爪哇岛（Java Island）命名。

光照需求：

养护难度：

Distribution: Malaysia and Indonesia (Java Island).

Morphology: Subshrub. Stem woody, ca. 4mm diameter. Leaf blade thin papyraceous, oblong, ca. 11cm long and ca. 5cm wide, margin undulant, apex acute, green, sometimes with silver spots, midvein distinct, lateral vein obscure. Umbel 20-40 flowered; flower ca. 1.6cm diameter, light lemon fragrance; corolla reflexed, corolla center white, lobe light yellow to yellow; corona white and protuberant, corolla base tomentose; stamen black.

Habits: Gradual increasing of light source help for growing of leaves and blossom of this *Hoya*; drought-sensitive, regular irrigation needed; loose substrate preferred; mature branch cuttings form roots easily.

Note: This *Hoya* is comes from Java Island, Indonesiaand it was named afterthis place.

Lighting requirement:

Maintenance difficulty:

Hoya javanica 'Alba'

白花爪哇球兰

形态特征：本栽培种为原种的白花品种，2010年发现后经选育逐步稳定性状；其与原种的区别在于花冠全部白色，并具浓香气。

栽培习性：同原种。

光照需求：☀ ☀ ☀

养护难度：🌱 🌱

Morphology: White flower; after breeding constant characters found since 2010; different between original species and this cultivar corolla is entirely white and strong fragrance.

Habits: Same as original species.

Lighting requirement: ☀ ☀ ☀

Maintenance difficulty: 🌱 🌱

Hoya kanyakumariana A. N. Henry & Swamin.in J. Bombay Nat. Hist. Soc. **75(2)**: 462. 1978.

印南球兰（新拟） 堪雅酷玛瑞球兰

分布：印度（南部）。

形态特征：悬垂藤本。茎直径约 2mm。叶片略肉质，硬，圆形至倒卵圆形，直径约 2cm，边缘波状，绿色，有时具银色斑点，叶脉不显。伞形花序呈球状，具花 8~15 朵；花直径约 1.5cm，具甜香气；花冠平展，先端略向背面反卷；白色，被短茸毛；副花冠白色；雄蕊深紫红色。

栽培习性：喜光；耐旱，但不耐水湿，在干燥的环境下生长良好，潮湿环境下易烂根，故需控制浇水频率；扦插枝条成活率高。

备注：本种以其产地印度南部的根尼亚古马里（Kanyakumari）命名。

光照需求：

养护难度：

Distribution: India (southern region).

Morphology: Hanging vine. Stem ca. 2mm diameter. Leaf blade rigid, fleshy, round to obovate, ca. 2cm diameter, margin undulant, green, sometimes with silver spots, vein obscure. Umbel convex, 8-15 flowered; flower ca. 1.5cm diameter, sweet fragrance; corolla spreading, apex slightly revolute, white tomentose; corona white; stamen deep fuchsia.

Habits: Plant likes bright environment; drought-resistant; dislikes water-logging; grows well in dry environment; roots tend to rot in moist environment, controlled water irrigation need; cuttings have high survival rate.

Note: This *Hoya* comes from Kanyakumari, Tamil Nadu, southern India and it was named after this place.

Lighting requirement:

Maintenance difficulty:

Hoya kastbergii Kloppenb.in Fraterna **16**(4): 1. 2003.

卡斯堡球兰

区域：马来西亚（婆罗洲岛）。

形态特征：攀援藤本。茎直径约 3mm。叶片厚革质，圆形，直径约 5cm，先端钝，淡绿色，表面光滑，叶脉不显。伞形花序具花 10~18 朵；花直径约 1cm，具甜香气；花冠平展，黄色，被白色短茸毛；副花冠白色至粉红色，雄蕊紫红色。

栽培习性：喜明亮的散射光；喜高温、高湿且通风的环境；栽培需选用疏松、透气的基质，忌水湿，亦忌干燥，故需保持基质干湿适中；扦插枝条生根慢，成活率较低；成活的插条生长较快，健壮植株枝叶茂盛。

备注：本种以瑞典植物采集人 Anne Kasterberg 命名。

光照需求：

养护难度：

Distribution: Malaysia (Borneo Island)

Morphology: Climbing vine. Stem ca. 3mm diameter. Leaf blade leathery, round, ca. 5cm diameter, apex obtuse, light green, smooth surface, vein obscure. Umbel 10-18 flowered; flower ca. 1cm diameter, sweet fragrance; corolla spreading, yellow, tomentose; corona white to pink; stamen fuchsia.

Habits: Plant likes bright scattered light, warm and moist environment; loose and ventilate substrate preferred; avoid water-logging and drought, and keep the substrate in moderate humidity; cuttings form roots slowly, and low survival rate; cuttings grow fast; healthy and mature plant has lush branches and leaves.

Note: Named after the Swedish collector, Anne Kasterberg.

Lighting requirement:

Maintenance difficulty:

Hoya kentiana C. M. Burton in Hoyan, **12**(3–2): IV. 1991.

冰糖球兰

分布：菲律宾。

形态特征：悬垂藤本，茎直径约 2mm。叶片肉质，披针形，长约 12cm，宽约 1.5cm，边缘全缘，先端渐尖，下面浅绿色，上面深绿色，常具紫色边，叶脉不显。伞形花序具花 12~25 朵；花直径约 1.2cm，具淡甜香气；花冠初时平展，后向背面强烈反卷呈球形，红色至紫红色，表面被短茸毛；副花冠红色至紫红色，常具黄色斑点；雄蕊黄色。

栽培习性：喜半阴；喜温暖、潮湿的环境，不耐寒，10℃以下易受寒害，冬季避免吹风；栽培需选用保湿但不积水并透气的基质；扦插宜采用成熟枝条。

备注：本种以英国园艺工作者 Douglas Kent 命名。

光照需求：☀☀☀

养护难度：🌱🌱🌱

Distribution: Philippines.

Morphology: Hanging vine. Stem ca. 2mm diameter. Leaf blade fleshy, lanceolate, ca. 12cm long and ca. 1.5cm wide, margin entire, apex acuminate, abaxially light green, adaxially deep green, margin usually purple, vein obscure. Umbel 12-25 flowered; flower ca. 1.2cm diameter, sweet fragrance; corolla spreading at early stage, and then strong revolute in ball shape, red to fuchsia, tomentose; corona red to fuchsia, often with yellow spots; stamen yellow.

Habits: Grows well in half-shading, warm and moist environment; nonhardy, tends to suffer from chilling temperature drops below 10 C ; avoid draught in cold season; moist and non-water-logging substrate preferred; mature branches are better for cuttings.

Note: Named after the British gardener, Douglas Kent.

Lighting requirement: ☀☀☀

Maintenance difficulty: 🌱🌱🌱

Hoya kerrii Craib in Bull. Misc. Inform. Kew, 1911(10): 418–419. 1911.

凹叶球兰 心叶球兰

分布：缅甸、老挝、泰国、越南和马来西亚。

形态特征：攀援藤本，茎直径 6~8mm。叶片肉质，倒心形，长约 10cm，宽约 7cm，先端下凹，绿色，表面光滑。伞形花序呈球状，具花 13~25 朵；花直径约 1cm，有淡酸气；花冠初期平展，后向背面反卷，淡黄白色，后变为淡黄色至浅褐色；副花冠红色。

栽培习性：喜光，光照充足则叶片厚实且形状饱满；较耐寒；如水分充足则生长更旺盛，扦插枝条易成活，叶片扦插也能生根，但较难萌发新芽。

备注：本种以植物学家 Arthur Francis George Kerr 命名。

光照需求：●●●●

养护难度：🌱

Distribution: Myanmar, Laos, Thailand, Vietnam and Malaysia.

Morphology: Climbing vine. Stem 6-8mm diameter. Leaf blade fleshy, obcordate, ca. 10cm long and ca. 7cm wide, apex sinus, green, surface smooth. Umbel 13-25 flowered; flower ca. 1cm diameter, light sourish odour; corolla spreading at early stage, then revolute, light yellowish white, then become yellow to light brown; corona red.

Habits: Plant likes bright environment; leaf blade thick; cold-resistant; grows lusher in proper irrigation; cuttings have high survival rate; leaf cuttings also form roots, but difficult to regenerate new buds.

Note: Named after the botanist Arthur Francis George Kerr.

Lighting requirement: ●●●●

Maintenance difficulty: 🌱

Hoya kerrii 'Albo-marginata'

金边凹叶球兰 心叶球兰"外锦"

形态特征：本栽培种与原种的区别在于叶边缘淡黄色至黄色。

栽培习性：同原种。

光照需求：☀-☀-☀-☀

养护难度：🔧

Morphology: This cultivar has light yellow to yellow leaf margin.

Habits: Same as original species.

Lighting requirement: ☀-☀-☀-☀

Maintenance difficulty: 🔧

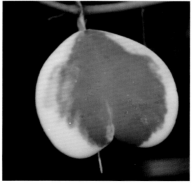

Hoya kerrii 'Spot leaf'

银斑凹叶球兰 银斑心叶球兰

形态特征：本栽培种与原种的区别在于叶片满布银色斑点。

光照需求：☀ ☀ ☀ ☀

养护难度：🌱

Morphology: This cultivar has silver spots all-over the leaf surface.

Habits: Same as original species.

Lighting requirement: ☀ ☀ ☀ ☀

Maintenance difficulty: 🌱

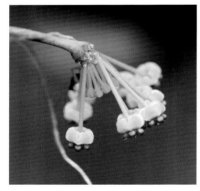

Hoya kerrii 'Variegata green'

金心凹叶球兰 心叶球兰"内锦"

形态特征：本栽培种与原种的区别在于叶中间具黄绿色、淡黄色至黄色斑块。

栽培习性：同原种。

光照需求：☀☀☀☀☀

养护难度：🔧

Morphology: This cultivar has yellowish green, light yellow to yellow spots on leaf center.

Habits: Same as original specise.

Lighting requirement: ☀☀☀☀☀

Maintenance difficulty: 🔧

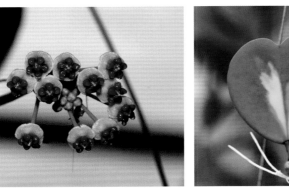

Hoya kloppenburgii T. Green in Fraterna, 14(2): 11. 2001.

柯氏球兰

分布：加里曼丹岛。

形态特征：攀援藤本。茎直径约4mm。叶片肉质，硬，阔披针形，长约11cm，宽约5cm，边缘全缘，先端渐尖，下面紫红色，上面通常深绿色，环境光照较强及昼夜温差较大时变为深红色，表面有时具银色斑点。伞形花序具花10~15朵；花直径约1.4cm，具清香气；花冠中部以下平展，中部以上向背面强烈反卷，黄色，表面被短茸毛；副花冠浅黄色。

栽培习性：喜明亮的散射光；喜温暖、潮湿的环境；扦插宜选择成熟枝条，较易生根，但新芽萌发缓慢；植株生长缓慢；新生枝条需长至很长才能萌发新叶。

备注：本种以园艺学家及采集人 Dale Kloppenburg 命名。

光照需求：☀ ☀ ☀ ☀ ☀

养护难度：🔧 🔧

Distribution: Kalimantan Island.

Morphology: Climbing vine. Stem ca. 4mm diameter. Leaf blade fleshy, rigid, lanceolate, ca. 11cm long and ca. 5cm wide, margin entire, apex acuminate, abaxially fuchsia, adaxially deep green; usually turn to deep red in strong light or temperature fluctuation between day and night, silver spots on the surface. Umbel 10-15 flowered; flower ca. 1.4cm diameter, light fragrance; corolla lower part spreading, upper part greatly revolute, yellow, tomentose; corona light yellow.

Habits: Plant likes bright scattered light, and grows well in warm and moist environment; mature branches are better for cuttings; cuttings form roots easily, but slow in new buds formation; slowly-growing; young branches can only produce new buds when it grow very long.

Note: Named after the horticulturist and collector, Dale Kloppenburg.

Lighting requirement: ☀ ☀ ☀ ☀ ☀

Maintenance difficulty: 🔧 🔧

Hoya lacunosa Blume, Bijdr. Fl. Ned. Ind. 16: 1063. 1826.

裂瓣球兰

分布：老挝、泰国、马来西亚和印度尼西亚（加里曼丹岛、苏门答腊岛和爪哇岛）。

形态特征：悬垂藤本。茎直径约 2mm，肉质。叶片卵形至阔披针形，长约 5cm，宽约 2.5cm，边缘全缘，先端急尖，下面浅绿色，上面深绿色，叶脉常浅绿色，突起。伞形花序呈平头状，具花 20~30 朵；花直径约 5mm，具浓香气，夜晚更甚；花冠向背面强烈反卷呈球状，白色，密被绢毛；副花冠白色，边缘宽大，雄蕊黄色。

栽培习性：喜阴，需避免阳光直射；根不耐水湿，故栽培宜选用疏松、透气的基质，忌积水；选用成熟枝条扦插易生根；植株生长较慢，但健壮后开花较多。

光照需求：☀ ☀

养护难度：🔧

Distribution: Laos, Thailand, Malaysia and Indonesia (Kalimantan Island, Sumatra Island and Java Island).

Morphology: Hanging vine. Stem ca. 2mm diameter, fleshy. Leaf blade ovate to wide lanceolate, ca. 5cm long and ca. 2.5cm wide, margin entire, apex acute, abaxially light green, adaxially deep green, vein light green, convex. Umbel flat-topped, 20-30 flowered; flower ca. 5mm diameter, strong fragrance, and becomes stronger at night; corolla strong revolute in ball shape, white, densely sericeous; corona white, margin wide; stamen yellow.

Habits: Plant likes shade; avoid direct sun light; roots are not moist-resistant, loose and ventilate substrate preferred; mature branch cuttings form roots easily; slowly-growing, blooms lush when it grows maturely.

Lighting requirement: ☀ ☀

Maintenance difficulty: 🔧

Hoya lambii T. Green in Fraterna, **13**(2): 2. 2000.

兰氏球兰

分布：加里曼丹岛。

形态特征：攀援藤本。茎直径约 5mm，常木质化。叶片纸质，较硬，卵圆形至倒卵形，宽大，长约 13cm，宽约 9cm，先端具小尖头，表面略粗糙，浅绿色，叶脉明显。伞形花序呈球状，具花 30~40 朵；花直径约 2cm，具淡柑橘香气；花冠内弯呈爪状，黄绿色，外面常具红褐色斑点；副花冠淡黄白色至黄绿色。

栽培习性：喜半阴；喜温暖、潮湿的环境；扦插宜选用成熟枝条；植株生长较慢；本种的叶在植株基部密生，向上逐渐稀疏；健壮植株开花较频繁。

备注：本种以植物学家 Anthony Lamb 命名。

光照需求：

养护难度：

Distribution: Kalimantan Island.

Morphology: Climbing vine. Stem ca. 5mm diameter, often woody. Leaf blade papyraceous, rigid, ovate to obovate, wide, ca. 13cm long and ca. 9cm wide, apex mucronulate, adaxially rough, light green, vein distinct. Umbel convex, 30-40 flowered; flower ca. 2cm diameter, light citrus fragrance; corolla incurved, claw-like, yellowish green, often red brown spots on the outer surface; corona light yellowish white to yellowish green.

Habits: Grows well in half-shading, warm and moist environment; mature branches are better for cuttings; slowly-growing; dense leaves at the base of the plant, and become sparse at upward; healthy plant blooms frequently.

Note: Named after the botanist, Anthon Lamb.

Lighting requirement:

Maintenance difficulty:

Hoya lanceolata Wallich ex D. Don, Prodr. Fl. Nepal. 130. 1825.

亚贝球兰

分布：不丹、印度、尼泊尔和缅甸。

形态特征：悬垂藤本。茎直径约 3mm。叶片肉质，较柔软，披针形，长约 2.5cm，宽约 1.5cm，边缘波状，先端渐尖，绿色。伞形花序呈平头状，具花 7~12 朵，排列整齐；花直径约 2cm，具淡香气；花冠平展，花冠裂片先端有时略向背面反卷，白色；副花冠粉红色，略透明。

栽培习性：喜阴；在通风、凉爽的环境下生长较佳；栽培宜选用疏松、透气且利于根部吸附的基质；成熟枝条扦插较易生根；植株生长较缓慢。

光照需求：☀☀

养护难度：🌱🌱🌱

Distribution: Bhutan, India, Nepal and Myanmar.

Morphology: Hanging vine. Stem ca. 3mm diameter. Leaf blade fleshy, soft, lanceolate, ca. 2.5cm long and ca. 1.5cm wide, margin undulant, apex acuminate, green. Umbel flat-topped, 7-12 flowered; flower neatly arranged, ca. 2cm diameter, light fragrance; corolla spreading, lobe apex sometimes revolute, white; corona pink, slightly transparent.

Habits: Plant likes shade and grows well in ventilate and cool environment; prefers loose and ventilate substrate to help the roots attach; mature branch cuttings form roots easily; slowly-growing.

Lighting requirement: ☀☀

Maintenance difficulty: 🌱🌱🌱

Hoya lasiantha Korthals ex Blume, Rumphia, **4**: 30. 1849.

棉毛球兰

分布：马来西亚和印度尼西亚（加里曼丹岛）。

形态特征：灌状藤本。茎直径约 5mm。叶片椭圆形，纸质，长约11cm，宽约6cm，边缘全缘，先端急尖，绿色，上面有时具银色斑点，叶脉清晰，在下面突起。伞形花序下垂，具花 7~15 朵；花直径约 2cm，具淡香气；花冠向背面反折，橙黄色，内面基部具茸毛，向上变无毛，副花冠橙黄色至淡黄色。

栽培习性：喜半阴，光照过强叶片易晒伤形成褐色晒斑；喜温暖潮湿的环境；栽培宜选择利于根部吸附并保水透气的基质；成熟枝条扦插易成活；植株生长较快；本种花型奇特，是著名的观赏种类。

光照需求：☀☀☀☀

养护难度：🌱🌱

Distribution: Malaysia and Indonesia (Kalimantan Island).

Morphology: Shrub-like vine. Stem ca. 5mm diameter. Leaf blade ellipse, papyraceous, ca. 11cm long and ca. 6cm wide, margin entire, apex acute, green, sometimes silver spots on the surface; vein distinct, convex abaxial. Umbel pendulous, 7-15 flowered; flower ca. 2cm diameter, light fragrance; corolla reflexed, orange, inner base tomentose, upper smooth; corona orange to light yellow.

Habits: Plant likes half-shading, warm and moist environment; the leaf tend to have brown sun burns when suffer from strong sun light; ventilate and water-preserving substrate can help the roots for attachment; mature branch cuttings have high survival rate; fast-growing; becomes a famous ornamental species with its unique flower shape.

Lighting requirement: ☀☀☀☀

Maintenance difficulty: 🌱🌱

Hoya lasiogynostegia P. T. Li in Bull. Bot. Res., Harbin, 4(1): 118–120, pl. 1. 1984.

橙花球兰

分布：中国（海南）。

形态特征：悬垂藤本，茎直径约 3mm。叶片卵状披针形，长 5~7cm。宽约 2.5cm，基部圆，先端尾尖，浅绿色，偶具银色斑点，中脉在叶片上面平，侧脉不明显。伞形花序呈平头状，具花 7~12 朵；花直径约 1cm，具浓香气；花冠平展，浅黄色；副花冠黄褐色。

栽培习性：喜阴；喜温暖潮湿的环境；栽培宜选用疏松、透气的基质；扦插枝条不易生根，成活率较低；植株生长较缓慢。

光照需求：☀☀

养护难度：🛠🛠🛠🛠

Distribution: China (Hainan).

Morphology: Hanging vine. Stem ca. 3mm diameter. Leaf blade oval lanceolate, 5-7cm long and ca. 2.5cm wide, base round, apex caudate, light green, occasionally with silver spots; vein flat, lateral veins obscure. Umbel flat-topped, 7-12 flowered; flower ca. 1cm diameter, strong fragrance; corolla spreading, light yellow; corona yellow brown.

Habits: Plant likes shade, warm and moist environment; loose and ventilate substrate preferred; cuttings are difficult to form roots, low survival rate; slowly-growing.

Lighting requirement: ☀☀

Maintenance difficulty: 🛠🛠🛠🛠

Hoya leucorhoda Schltr. in Bot. Jahrb. Syst. **50**: 119. 1913.

白玫瑰红球兰

分布：巴布亚新几内亚。

形态特征：攀援藤本，茎直径约 2.5mm。叶片纸质，椭圆形，长约 8cm，宽约 5cm，边缘全缘，先端渐尖至尾尖，深绿色，光滑，叶脉明显。伞形花序具花 15~25 朵；花直径约 2cm，具淡香气；花冠平展或内弯呈爪状，白色，被短茸毛；副花冠浅黄色。

栽培习性：喜半阴；喜温暖潮湿的环境，环境湿度过低叶片易脱落，有一定的耐寒能力，可耐 5℃ 上下的低温约一个月；成熟枝条扦插易成活；植株生长较快；容易开花。

光照需求：☀☀☀

养护难度：🌱🌱

Distribution: Papua New Guinea.

Morphology: Climbing vine. Stem ca. 2.5mm diameter. Leaf blade papyraceous, ellipse, ca. 8cm long and ca. 5cm wide, margin entire, apex acuminate to caudate, deep green, surface smooth, vein distinct. Umbel 15-25 flowered; flowerca. 2cm diameter, light fragrance; corolla spreading or incurved and claw-like, white, tomentose; corona light yellow.

Habits: Plant grows well in half-shading, warm and moist environment; leaves tend to drop in low humidity; cold-resistance, can endure a low temperature (5 ℃)for one month; mature branch cuttings have high survival rate; plant grows fast and blooms early.

Lighting requirement: ☀☀☀

Maintenance difficulty: 🌱🌱

Hoya limoniaca S. Moore in J. Linn. Soc., Bot. 45: 368. 1921.

黎檬球兰

分布：新喀里多尼亚。

形态特征：攀援藤本。茎直径约 2mm。叶片较薄，柔软，卵形至卵状披针形，长约 6cm，宽 3~4cm，边缘全缘，先端渐尖至急尖，深绿色，偶见银色斑点，表面光滑，叶脉清晰。伞形花序呈球状，具花 16~20 朵；花直径约 1.8cm，具淡香气；花冠初期平展，后期略向背面反折，白色至淡黄白色；副花冠白色；雄蕊红色。

栽培习性：适应性强；对光照要求不高，明亮或半阴的环境下栽培均可；宜选择透气且保湿的栽培基质；成熟枝条扦插较易生根；植株生长较快。

光照需求：
养护难度：

Distribution: New Caledonia.

Morphology: Climbing vine. Stem ca. 2mm diameter. Leaf blade thin, soft, ovate to ovate lanceolate, ca. 6cm long and 3-4cm wide, margin entire, apex acuminate to acute, deep green, occasionally silver spots, smooth, vein distinct. Umbel convex, 16-20 flowered; flowerca. 1.8cm diameter, light fragrance; corolla spreading at early stage, then slightly reflexed, white to light yellowish white; corona white; stamen red.

Habits: Well-adapted; bright or half-shading environment preferred; ventilate and moist substrate preferred; mature branch cuttings form roots easily; fast-growing.

Lighting requirement:
Maintenance difficulty:

Hoya linearis Wall.ex D. Don, Prodr. Fl. Nepal. 130. 1825.

线叶球兰

分布：中国（云南）、印度和尼泊尔。

形态特征：悬垂藤本。茎直径约 2mm。叶片近圆柱形，肉质，长3~5cm，宽约5mm，先端具小尖头，浅绿色，表面被短茸毛，叶脉不显。伞形花序下垂，呈平头状，具花10~15朵；花直径约7mm，具柠檬香气，夜晚更甚；花冠平展，裂片先端略向背面反卷，白色，表面密被短茸毛；副花冠白色，具粉红色斑点，略透明。

栽培习性：喜半阴，避免阳光直射；由于根系瘦弱，栽培需选择细颗粒的基质，基质不宜潮湿，应适当保持干爽；本种扦插枝条生根慢，成活率较低。

光照需求：

养护难度：

Distribution: China (Yunnan), India and Nepal.

Morphology: Hanging vine. Stem ca. 2mm diameter. Leaf blade fleshy, terete, 3-5cm long and ca. 5mm wide, apex mucronulate, light green, tomentose, vein obscure. Umbel pendulous, flat-topped, 10-15 flowered; flower ca. 7mm diameter, lemon fragrance, becomes stronger at night; corolla spreading, lobe apex slightly revolute, white, densely tomentose; white corona, with pink spots, slightly transparent.

Habits: Plant likes half shade; avoid direct sun light; this Hoya's roots are weak, so small-sized and undamped substrate preferred; keep the substrate in proper dryness; cuttings grow slowly, low survival rate.

Lighting requirement:

Maintenance difficulty:

Hoya lobbii J. D. Hook. Fl. Brit. India, 4(10): 54. 1883.

罗比球兰

分布：印度（北部）、缅甸和泰国。

形态特征：半灌木，茎直径约 4mm。叶片薄革质，阔披针形至椭圆形，长约10cm，宽4~5cm，边缘全缘，先端渐尖至急尖，绿色，偶具银色斑点，中脉在下面隆起，上面凹陷，侧脉明显。伞形花序具花10~14朵；花直径约2cm，几无香气；花冠向背面反折，粉红色、红色至紫红色，表面被短茸毛，副花冠红色至紫红色。

栽培习性：喜半阴；喜温暖、潮湿的环境；栽培宜选择疏松、透气且保湿的基质，基质积水则容易烂根；成熟枝条扦插易生根；适当增加施肥可促进花序分化；本种的花序一般生于枝条先端，故待植株生长健壮后可适当修剪以增加分枝，从而增加花序数量。

备注：本种以英国植物学家 Thomas Lobb 命名。

光照需求：

养护难度：

Distribution: India (northern region), Myanmar and Thailand.

Morphology: Subshrub. Stem ca. 4mm diameter. Leaf blade thin leathery, wide lanceolate to ellipse, ca. 10cm long and 4-5cm wide, margin entire, apex acuminate to acute, green, occasionally with silver spots; midvein convex on lower surface and concave on upper surface; lateral vein distinct. Umbel 10-14 flowered; flower ca. 2cm diameter, nearly odourless; corolla reflexed, red to fuchsia, surface tomentose; corona red to fuchsia.

Habits: Grows well in half-shading, warm and moist environment; loose ventilate and water-preserving substrate preferred; water-logging substrate may cause rotten roots; mature branch cuttings form roots easily; proper addition in fertilizers can help inflorescence grow; inflorescence usually on the upper part of branch, so proper pruning can help branching and forming more inflorescence.

Note: Named after the British botanist, Thomas Lobb.

Lighting requirement:

Maintenance difficulty:

Hoya loheri Kloppenb.in Fraterna 1991(3), Philipp. Hoya Sp. Suppl.: III. 1991.

洛黑球兰

分布：菲律宾。

形态特征：攀援藤本，茎直径约1mm。叶片肉质，厚，披针形至倒披针形，长约6cm，宽约1cm，边缘常向下面反卷，先端钝，下面淡绿色，常具红色斑块，上面绿色，叶脉不显。伞形花序具花15~25朵；花直径约8mm，具清香气；花冠向背面反卷呈扁球状，棕红色，被短茸毛；副花冠黄色；雄蕊红色。

栽培习性：喜半阴；喜温暖潮湿的环境，环境湿度较大则叶边缘平展；栽培宜选择保水、透气的基质；扦插枝条较难成活；本种株形较凌乱，需多修剪；健壮植株开花多。

备注：本种以德国植物采集人August Loher命名。

光照需求：☀☀☀

养护难度：🌱🌱🌱🌱

Distribution: Philippines.

Morphology: Climbing vine. Stem ca. 1mm diameter. Leaf blade fleshy, thick, lanceolate to oblanceolate, ca. 6cm long and ca. 1cm wide, margin revolute, apex obtuse, abaxially light green, with red spots; adaxially green, vein obscure. Umbel 15-25 flowered; flower ca. 8mm diameter, fragrance; corolla revolute in a flat ball shape, brownish red, tomentose; corona yellow; stamen red.

Habits: Plant likes half shade, warm and moist environment; leaf margin is spreading in high humidity; water-reserving and ventilate substrate preferred; cuttings are difficult to survive; plant shape is little messy, frequent trimming needed; healthy plant can bloom more flowers.

Note: Named after the German plant collector, August Loher.

Lighting requirement: ☀☀☀

Maintenance difficulty: 🌱🌱🌱🌱

Hoya longifolia Wall.ex Wight in Contr. Bot. India, 36. 1834.

长叶球兰

分布：中国（云南）、巴基斯坦、印度、不丹、尼泊尔和泰国。

形态特征：攀援藤本，茎直径约 2mm。叶片肉质，厚，条形至倒披针形，长约 8cm，宽 2~3cm，边缘全缘，先端尾尖，下面淡绿色，上面绿色。伞形花序具花 15~22 朵；花直径约 1.5cm，具浓香气；花冠平展或内弯呈爪状，白色至粉红色，被短茸毛；副花冠白色；雄蕊褐色、红褐色至紫红色。

栽培习性：喜半阴；对生长环境要求不高，略干燥至湿润的环境均可正常生长；成熟枝条扦插成活率高；植株生长快但攀附能力较差，需适当扶持；昼夜温差略大可促进开花。

光照需求：

养护难度：

Distribution: China (Yunnan), Pakistan, India, Bhutan, Nepal and Thailand.

Morphology: Climbing vine. Stem ca. 2mm diameter. Leaf blade fleshy, thick, linear to oblanceolate, ca. 8cm long and 2-3cm wide, margin entire, apex sharp ended, leaf back light green, leaf surface green. Umbel 15-22 flowered; flower ca. 1.5cm diameter, strong fragrance; corolla spreading or incurved and claw-like, white to pink, tomentose; corona white; stamen brown, red brown to fuchsia.

Habits: Plant likes half shade; low demand in environment; grows well in dry to moist environment; cuttings have high survival rate; fast-growing, not good at climbing, proper support needed; fluctuation temperature between day and night can help for plant blooming.

Lighting requirement:

Maintenance difficulty:

Hoya lucardenasiana Kloppenb., Siar & Cajano Asia Life Sci. 18(1): 144. 2009.

洛斯球兰（新拟）卢卡球兰

分布： 菲律宾（吕宋岛）。

形态特征： 攀援藤本。茎直径约 2mm。叶片革质，卵形，略向背面弯弓，长约 6cm，宽约 4cm，边缘全缘，先端急尖，绿色，叶脉不明显。伞形花序略呈球状，具花 15~25 朵；花直径约 1cm，具淡香气；花冠向背面反卷，粉紫色至红色，被短茸毛；副花冠深红色。

栽培习性： 适应性强；喜半阴；喜潮湿且通风的环境；栽培宜选用疏松、透气的基质，利于根部发育；扦插枝条易生根；植株生长快，可以适当增加施肥以支持生长；本种花序萌生早，开花频繁。

备注： 本种以菲律宾植物学家 Lourdes B. Cardenas 命名。

光照需求： ☀☀☀

养护难度： 🔧

Distribution: Philippines (Luzon Island).

Morphology: Climbing vine. Stem ca. 2mm diameter. Leaf blade leathery, ovate, slightly revolute, ca. 6cm long and ca. 4cm wide, margin entire, apex acute, green, vein obscure. Umbel convex, 15-25 flowered; flower ca. 1cm diameter, light fragrance; corolla revolute, pale lilac to red, tomentose; corona deep red.

Habits: Well-adapted and likes half-shading, moist and ventilate environment; loose and ventilate substrate can promote root formation; cuttings have high survival rate; fast-growing; proper additional fertilizers can help its growing; form inflorescence early, and blooms frequently.

Note: Named after the Philippine botanist, Lourdes B. Cardenas.

Lighting requirement: ☀☀☀

Maintenance difficulty: 🔧

Hoya lyi H. Lév. in Bull. Soc. Bot. France, 54(6): 369–370. 1907.

香花球兰

分布：中国（广西、贵州、四川和云南）。

形态特征：攀援藤本，茎直径约 3mm。叶片革质，椭圆形至椭圆状披针形，长约 7cm，宽约 2.5cm，边缘全缘，先端钝，深绿色，被短茸毛，叶脉不明显。伞形花序具花 8~16 朵；花直径约 1.8cm，具清香气；花冠平展，先端略向背面反卷，白色，被短茸毛；副花冠淡黄白色。

栽培习性：喜半阴；耐寒，可耐受 0℃左右的低温；栽培选择利于根吸附的基质则生长更佳；扦插采用成熟的枝条容易生根；根系生长稳定后植株生长速度很快。本种为我国特有种，模式标本产自贵州，其在我国栽培适应性强，花有宜人清香，是极佳的观赏种类。

光照需求：
养护难度：

Distribution: China (Guangxi, Guizhou, Sichuan and Yunnan).

Morphology: Climbing vine. Stem ca. 3mm diameter. Leaf blade leathery, ellipse to ellipse lanceolate, ca. 7cm long and ca. 2.5cm wide, margin entire, apex obtuse, deep green, tomentose, vein obscuree. Umbel 8-16 flowered; flower ca. 1.8cm diameter, fragrance; corolla spreading, lobe apex slightly revolute, white, tomentose; corona light yellowish white.

Habits: Plant likes half shade; cold-resistant, can endure a low temperature at 0 ℃ ; loose and ventilate substrate can help for roots attachment; mature branch cuttings form roots easily; fast-growing after roots grow formation; Chinese endemic species, type specimen was collected from Guizhou; well-adapted in China; pleasant fragrance flowers makes an excellent ornamental plant.

Lighting requirement:
Maintenance difficulty:

Hoya macgillivrayi F. M. Bailey in Queensland Agric. J. **1**: 190. 1914.

麦氏球兰（新拟） 麦季理斐球兰

分布：澳大利亚（昆士兰）。

形态特征：攀援藤本，茎直径约 3mm。叶片革质，硬，椭圆形至阔椭圆形，长约 10cm，宽约 6cm，基部心形，边缘全缘，先端具小尖头，深绿色，中脉明显，侧脉不显。伞形花序具花 4~8 朵；花直径约 7cm，具淡香气；花冠呈阔钟状，裂片先端尖，深红色至紫红色，里面基部颜色常较淡；副花冠深红色，雄蕊白色。

栽培习性：喜明亮的散射光；栽培宜选择利于植株根部吸附的基质；成熟枝条扦插成活率高；本种植株较大，栽培需预留足够空间；需待植株生长健壮后才能开花，保持栽培基质干燥—湿润—干燥的循环可促进开花。

备注：本种以博物学家 William MacGillivray 命名。

光照需求：☀☀☀☀☀

养护难度：🌱🌱

Distribution: Australia (Queensland).

Morphology: Climbing vine. Stem ca. 3mm diameter. Leaf blade leathery, rigid, ellipse to wide ellipse, ca. 10cm long and ca. 6cm wide, leaf base cordate, margin entire, apex mucronulate, deep green, midvein distinct, lateral vein obscure. Umbel 4-8 flowered; flower ca. 7cm diameter, light fragrance; corolla wide campanulate, lobe apex sharp, deep red to fuchsia; inner base usually lighter; corona deep red; stamen white.

Habits: Plant likes bright scattered light; loose and ventilate substrate can for help the roots attachment; mature branch cuttings have high survival rate; due to large size of the plant, large space should be given; bloom after maturation; keep the substrate in drying and watering cycle will promote its blooming.

Note: Named after the naturalist William MacGillivray.

Lighting requirement: ☀☀☀☀☀

Maintenance difficulty: 🌱🌱

Hoya megalaster Warb.ex K. Schum. & Lauterb. Fl. Schutzgeb. Südsee, 513. 1905.

红花球兰

分布：巴布亚新几内亚。

形态特征：攀援藤本，茎脆，直径约 2mm。叶肉质，椭圆形至阔椭圆形，长约 15cm，宽约 8cm，基部心形，边缘全缘，先端具小尖头，深绿色，叶脉在上面突起。伞形花序具花 8~12 朵；花大，直径约 4cm，几无香气；花冠内弯呈爪状，裂片宽，外面粉白色至红色，里面红色；副花冠红色，雄蕊黄白色。

栽培习性：喜半阴；不耐寒，冬季气温在 10℃ 以下易受寒害，需适当保温；成熟枝条扦插易成活；植株生长较缓慢，新萌发枝条需生长至足够长才开始展叶；健壮植株开花频繁；本种花大且颜色鲜艳，是极佳的观赏种类。

光照需求：

养护难度：

Distribution: Papua New Guinea.

Morphology: Climbing vine, stem fragile, ca. 2mm diameter. Leaf blade fleshy, ellipse to wide ellipse, ca. 15cm long and ca 8cm wide, leaf base cordate, margin entire, apex mucronulate, deep green, adaxially vein convex. Umbel 8-12 flowered; flower 4-5cm diameter, nearly odourless; corolla incurved, claw-like, lobe broad, outer surfacelight pink to red, inner surface red; corona red; stamen yellowish white.

Habits: Plant likes half shade; nonhardy, tends to suffer from chilling injury when temperature drops below to 10℃, proper insulation needed in cold seasons; mature branch cuttings have high survival rate; slowly-growing; healthy plant blooms frequently; large size and bright color make an excellent ornamental plant.

Lighting requirement:

Maintenance difficulty:

Hoya meliflua Merr. Sp. Blancoan. 318. 1918.

美丽球兰

分布： 菲律宾。

形态特征： 攀援藤本。茎肉质，直径约5mm 叶片肉质，长椭圆形至倒披针形，长11~13cm，宽约5~6cm，边缘全缘，先端渐尖至急尖，下面浅绿色，上面深绿色，叶脉不明显。伞形花序具花15~20朵；花直径约1.5cm，具清香气；花冠向背面反卷，红棕色、粉红色至紫红色，密被短茸毛；副花冠中心紫红色，边缘淡黄色。

栽培习性： 喜明亮的散射光；扦插宜选用成熟的枝条，插条生根快；植株生长较快；本种枝条节间长，生长需要较大空间；昼夜温差略大可促进开花。

光照需求： ※-※-※-※-※
养护难度： 🛠

Distribution: Philippines.

Morphology: Climbing vine. Stem fleshy, ca. 5mm diameter. Leaf blade fleshy, oblong to oblanceolate, 11-13cm long and 5-6cm wide, margin entire, apex acuminate to acute, abaxially light green, adaxially deep green, vein obscure. Umbel 15-20 flowered; flower ca. 1.5cm diameter, light fragrance; corolla revolute, red brown, pink to fuchsia, densely tomentose; corona center fuchsia red, margin light yellow.

Habits: Plant likes bright scattered light; mature branch cuttings have high survival rate; cuttings form roots fast; fast-growing; stem internode long, larger growing space needed; high temperature difference between day and night promote blooming.

Lighting requirement: ※-※-※-※-※
Maintenance difficulty: 🛠

Hoya merrillii Schltr. in Fragm. Fl. Philipp. **1**: 131. 1904.

美林球兰（新拟） 玛丽球兰

分布：菲律宾。

形态特征：攀援藤本，茎直径约3mm。叶片革质，长心形，长约10cm，宽约7cm，边缘全缘，先端急尖，通常绿色，有时橙红色，叶脉在上面突起，明显。伞形花序呈球状，具花20~30朵；花直径约1.2cm，具淡柠檬香气；花冠平展或略内弯，白色至淡黄色；副花冠白色；雄蕊浅黄色。

栽培习性：喜半阴，光照过强则新叶容易灼伤；栽培宜选用疏松且保湿的基质；植株生长较快，易养护；本种叶片在春季及秋冬光照略强的环境下变为鲜艳的橙红色或红色，观赏价值高。

备注：本种以美国植物学家Elmer Drew Merrill 命名。

光照需求：

养护难度：

Distribution: Philippines.

Morphology: Climbing vine. Stem ca. 3mm diameter. Leaf blade leathery, long cordate, ca. 10cm long and ca. 7cm wide, margin entire, apex acute, green, sometimes orange, vein convex, distinct. Umbel convex, 20-30 flowered; flower ca. 1.2cm diameter, light lemon fragrance; corolla spreading or slightly incurved, white to light yellow; corona white; stamen light yellow.

Habits: Plant likes half shade; young leaf easily burnt in strong sun light; loose and water-preserving substrate preferred; fast-growing and easy-cultivation; leaf turns orange or red in strong sun light (in spring, autumn and winter season), having great ornamental values.

Note: Named after the American botanist, Elmer Drew Merrill.

Lighting requirement:

Maintenance difficulty:

Hoya micrantha J. D. Hook. Fl. Brit. India, 4(10): 55. 1883.

小花球兰

分布：缅甸、泰国、老挝和马来西亚。

形态特征：攀援藤本。茎直径约2.5mm。叶常3片轮生，叶片革质，硬，披针形至阔披针形，长约8cm，宽约3.5~4.5cm，边缘全缘，先端渐尖，绿色，表面略粗糙，叶脉不显。伞形花序具花10~18朵，花直径约5mm，几乎无香气；花冠向背面反卷，淡黄色，具紫色斑点，被短茸毛；副花冠淡黄色，边缘紫红色；雄蕊淡黄色至黄色。

栽培习性：适应性强，喜半阴，对栽培环境要求不高，可适应不同的栽培基质。成熟枝条扦插易成活。植株生长快，开花较迟，需等植株健壮后才会萌生花序。

光照需求：🌣🌣🌣

养护难度：🌱🌱

Distribution: Myanmar, Thailand, Laos and Malaysia.

Morphology: Climbing vine. Stem ca. 2.5mm diameter. Leaf 3 whorled, leaf blade leathery, rigid, lanceolate to wide lanceolate, ca. 8cm long and 3.5-4.5cm wide, margin entire, apex acuminate; green, surface rough, vein obscure. Umbel 10-18 flowered; flower ca. 5mm diameter, nearly odourless; corolla revolute, light yellow, purple spots, tomentose; corona light yellow, margin fuchsia; stamen light yellow to yellow.

Habits: Well-adapted and likes half shade; low demand in environment; adapted to many kinds of substrate; mature branch cuttings have high survival rate; fast-growing, blooms late, and produces inflorescence after maturation.

Lighting requirement: 🌣🌣🌣

Maintenance difficulty: 🌱🌱

Hoya mindorensis Schltr. in Philipp. J. Sci. 1(Suppl.): 303. 1906.

民都洛球兰（新拟） 棉德岛球兰

分布：菲律宾。

形态特征：攀援藤本。茎直径约 3mm。叶片革质至肉质，椭圆形，长约 10cm，宽 4~5cm，边缘全缘，先端渐尖，中脉明显，侧脉不明显。伞形花序呈球状，具花 20~35 朵；花直径约 1cm，几乎无香气；花冠初期内弯，后向背面反卷，黄色、红色至紫红色，被白色长绢毛；副花冠紫红色；雄蕊淡黄色。

栽培习性：喜半阴，不甚耐寒，冬季需适当保温。成熟枝条扦插易生根，植株生长快，开花频繁。本种在不通风环境下栽培易滋生吹棉介壳虫及蚜虫，应注意环境通风。

备注：本种以其产地菲律宾的民都洛岛（Mindoro Island）命名。

光照需求：

养护难度：

Distribution: Philippines.

Morphology: Climbing vine. Stem ca. 3mm diameter. Leaf blade leathery to fleshy, ellipse, ca. 10cm long and 4-5cm wide, margin entire, apex acuminate, midvein distinct, lateral vein obscure. Umbel convex, 20-35 flowered; flower ca. 1cm diameter, nearly odourless; corolla incurved at early stage, then revolute, yellow, red to fuchsia, white sericeous; corona fuchsia; stamen light yellow.

Habits: Plant likes half shade; nonhardy, needs proper insulation in winter; mature branch cuttings form roots easily; fast-growing, blooms frequently; tends to suffer from scale insect and aphid in unventilated environment.

Note: This *Hoya* comes from Mindoro Island, Philippines and it was named after this place.

Lighting requirement:

Maintenance difficulty:

Hoya multiflora Blume in Catalogus, 49. 1823.

蜂出巢 流星球兰

分布：中国（广西和云南）、缅甸、老挝、越南、泰国、菲律宾、马来西亚和印度尼西亚。

形态特征：半灌木。茎直径约5mm。叶片纸质，椭圆形，长约13cm，宽约6cm，深绿色，叶脉在上面凹陷，不明显。伞形花序具花25~40朵；花直径约2cm，几乎无香气；花冠向背面反折，中间白色，边缘黄色，或花冠全部黄色；副花冠白色；雄蕊红褐色。

栽培习性：适应性强，喜半阴，植株较耐寒，夏季温度过高则影响生长。扦插采用成熟的枝条，置于阴凉的环境下可快速生根抽芽。健壮植株全年可持续开花，适当修剪枝条可促进花序萌生，使开花量更大。

光照需求：

养护难度：

Distribution: China (Guangxi and Yunnan), Myanmar, Laos, Vietnam, Thailand, Philippines, Malaysia and Indonesia.

Morphology: Subshrub. Stem ca. 5mm diameter. Leaf blade papyraceous, ellipse, ca. 13cm long and ca. 6cm wide, deep green, vein obscure, concave on upper surface. Umbel 25-40 flowered; flower ca. 2cm diameter, nearly odourless; corolla reflexed, center white, margin yellow, or entirely yellow; corona white; stamen red brown.

Habits: Well-adapted and likes half shade; cold-resistant; high temperature may hinder growing; mature branches as cuttings form roots easily; form roots and buds fast in shade environment; healthy plant blooms throughout the year; proper pruning can promote inflorescence formation and increase plenty of flowers.

Lighting requirement:

Maintenance difficulty:

Hoya naumannii Schltr. in Bot. Jahrb. Syst. **40**(Beibl. 92): 15. 1908.

瑙曼球兰（新拟） 瑙珉球兰

分布：新几内亚岛。

形态特征：攀援藤本。茎直径约 5mm。叶片革质，阔椭圆形，长约 10cm，宽约 7cm，边缘全缘，先端尾尖，深绿色，叶脉明显。伞形花序具花 18~25 朵；花直径约 1.6cm，具玫瑰香气；花冠略内弯，粉紫色，边缘白色，或全部为白色；副花冠中间粉红色，边缘白色；雄蕊黄色。

栽培习性：喜半阴，温暖潮湿的气候。成熟并略木质化的枝条扦插易生根，插条生长快。早期即需扶持以获得美观的株形，植株较大，栽培需预留足够的生长空间。本种常年开花，是极佳的观赏种类。

备注：本种以其发现人 Carl Friedrich Naumann 命名。

光照需求：☀☀☀

养护难度：🌱🌱

Distribution: New Guinea Island.

Morphology: Climbing vine. Stem ca. 5mm diameter. Leaf blade leathery, wide ellipse, ca. 10cm long and ca. 7cm wide, margin entire, apex caudate, deep green, vein distinct. Umbel 18-25 flowered; flower ca. 1.6cm diameter, rose fragrance; corolla slightly incurved, pale lilac, corolla margin white, center pink, or corolla entirely white; stamen yellow.

Habits: Grows well in half-shading, warm and moist environment; mature and slight woody branch cuttings form roots easily; cuttings grow fast; proper support needed for beautiful plant shape; due to large size of the plant, large growing space should be given; plant blooms throughout the year and having great ornamental value.

Note: Named after its discoverer, Carl Friedrich Naumann.

Lighting requirement: ☀☀☀

Maintenance difficulty: 🌱🌱

Hoya neocaledonica Schltr. in Bot. Jahrb. Syst. **39**: 245. 1906.

新喀里多尼亚球兰（新拟）

分布： 新喀里多尼亚。

形态特征： 攀援藤本。茎柔软，直径约 2mm。叶略肉质，柔软，椭圆形，长约 10cm，宽 5~6cm，边缘全缘，先端渐尖，浅绿色，叶脉明显。伞形花序具花 10~15 朵；花直径约 1.2cm，具玫瑰香气；花冠向背面反折，粉红色；副花冠白色或淡粉色。

栽培习性： 喜阴，光照过强则抑制生长，喜通风、凉爽的环境，忌闷热。扦插枝条易成活，植株生长较快。本种花期在夏季，可多次开花，其他季节零星开花。

备注： 本种以其产地新喀里多尼亚（Nouvelle-Calédonie）命名。

光照需求： ☀☀

养护难度： 🌱🌱

Distribution: New Caledonia.

Morphology: Climbing vine. Stem soft, ca. 2mm diameter. Leaf blade fleshy, soft, ellipse, ca. 10cm long and 5-6cm wide, margin entire, apex acuminate, light green, vein distinct. Umbel 10-15 flowered; flower ca. 1.2cm diameter, rose fragrance; corolla reflexed, pink; corona white or light pink.

Habits: Plant likes shade; bright light hold up its growing; ventilation and coolness preferred; avoid stuffiness; cuttings have high survival rate; fast-growing; blooms several times in summer and blooms fragmentarily in other time.

Note: Named after its distributing area—Nouvelle-Calédonie (New Caledonia).

Lighting requirement: ☀☀

Maintenance difficulty: 🌱🌱

Hoya nicholsoniae F. Muell. Fragm. **5**: 159. 1866.

秋水仙球兰

分布：新几内亚岛、澳大利亚和波利尼西亚。

形态特征：攀援藤本。茎直径约3mm。叶片薄革质，阔披针形至长椭圆形，长约8cm，宽约4cm，边缘全缘，先端渐尖，深绿色，中脉及侧脉均明显。伞形花序呈球状，具花12~18朵；花直径约1.2cm，具淡香气；花冠向背面反折，黄绿色；副花冠白色。

栽培习性：适应性强，易养护。喜阴、喜温暖潮湿的环境。栽培选用疏松、透气的基质可促进根系发育，扦插枝条易生根。植株生长快，故在早期需进行必要的塑形以获得美观的形态，健壮植株终年开花不断，是养护容易且美观的种类。

光照需求：☀☀

养护难度：🔧

Distribution: New Guinea Island, Australia and Polynesia.

Morphology: Climbing vine. Stem ca. 3mm diameter. Leaf blade thin leathery, wide lanceolate to oblong, ca. 8cm long and ca. 4cm wide, margin entire, apex acuminate, deep green, midvein and lateral vein distinct. Umbel 12-18 flowered; flower ca. 1.2cm diameter, light fragrance; corolla reflexed, yellowish green; corona white.

Habits: Well-adapted and easy-cultivated; grows well in shading, warm and moist environment; loose and ventilate substrate help to form roots; cuttings have high survival rate; fast-growing, early stage plant shaping necessary to make it beautifully; healthy plant blooms throughout the year; maintainable and beautiful.

Lighting requirement: ☀☀

Maintenance difficulty: 🔧

Hoya nummularioides Costantin in Fl. Indo-Chine **4**: 129. 1912.

钱币球兰（新拟）钱叶球兰

分布：越南、老挝、泰国和柬埔寨。

形态特征：攀援藤本。茎直径约 3mm。叶片厚革质，卵形，长约 3cm，宽约 2cm，边缘加厚，深绿色，先端急尖，绿色，表面粗糙，被短茸毛，叶脉不显。伞形花序具花 15~25 朵；花直径约 6mm，具浓郁的甜香气；花冠平展，白色，被短茸毛；副花冠中间红色至粉红色，边缘白色。

栽培习性：喜明亮的散射光，光照不足则枝条徒长。喜通风且略干爽的环境，忌潮湿。栽培宜选择疏松、透气且易干的基质，基质长期潮湿根部发育不良，保持干湿循环植株生长较佳。扦插枝条易生根。本种花期为秋季，环境适宜则开花极多，花香浓郁。本种喜光，浇水间隔周期长，是美观且适合粗放养护的种类。

光照需求：☀☀☀☀☀

养护难度：🔧

Distribution: Vietnam, Laos, Thailand and Cambodia.

Morphology: Climbing vine. Stem ca. 3mm diameter. Leaf blade thick leathery, ovate, ca. 3cm long and ca. 2cm wide, margin thick, apex acute, green, surface rough, tomentose, vein obscure. Umbel 15-25 flowered; flower ca. 6mm diameter, strong sweet fragrance; corolla spreading, white, tomentose; corona center red to pink, margin white.

Habits: Plant likes bright scattered light; branches grow excessively when insufficient light; ventilate and a little dry environment preferred; avoid moisture; loose, ventilate and dry substrate preferred; moist substrate leads maldevelopment of the roots; keep drying and watering cycle help plant growth; cuttings have high survival rate; plant blooms in autumn; plenty of flowers give strong fragrance; plant likes bright light; long irrigation intervals; beautiful and maintainable.

Lighting requirement: ☀☀☀☀☀

Maintenance difficulty: 🔧

Hoya obovata Decne. in Prodr. 8: 635. 1844.

倒卵叶球兰 镜叶球兰

分布：印度、泰国、印度尼西亚和斐济。

形态特征：攀援藤本。茎直径约5mm。叶片肉质、硬，圆形至倒卵形，直径约8cm，边缘全缘，先端圆或微凹，深绿色，表面常具银色斑点，叶脉不显。伞形花序具花15~25朵；花直径约1.5cm，具甜香气；花冠平展或略向背面反折，白色至淡粉色，被短茸毛；副花冠红色至紫红色，边缘颜色较淡。

栽培习性：适应性强。喜半阴，在较强光照下亦能生长且叶片变为红褐色。成熟枝条扦插成活率高，插条新芽萌发时适当施肥能促进快速生长。栽培宜选用疏松、透气的基质，植株体型较大，需预留足够空间供其生长。本种养护简单，叶形奇特，常年开花不断，是优良的观赏种类。

光照需求：
养护难度：

Distribution: India, Thailand, Indonesia and Fiji.

Morphology: Climbing vine. Stem ca. 5mm diameter. Leaf blade fleshy, rigid, round to obovate, ca. 8cm diameter, margin entire, apex round or retuse, deep green, often silver spots on the surface, vein obscure. Umbel 15-25 flowered; flower ca. 1.5cm diameter, sweet fragrance; corolla spreading or slightly reflexed, white to light pink, tomentose; corona red to fuchsia, margin lighter.

Habits: Well-adapted and likes half shade; grow in strong light and leaf blade will turn red brown; mature branch cuttings have high survival rate; proper additional fertilizer can help the cuttings grow fast when producing new buds; loose and ventilate substrate preferred; due to large size of the plant, large growing space should be given; easy to cultivate; unique leaves and blooms throughout the year, having great ornamental value.

Lighting requirement:
Maintenance difficulty:

Hoya obscura Elmer ex C. M. Burton in Hoyan 8(1): 15. 1986.

小棉球球兰

分布：菲律宾。

形态特征：攀援藤本。茎直径约 3mm。叶片革质，阔披针形至椭圆形，长约 8cm，宽 4~5cm，边缘全缘，先端渐尖至急尖，绿色，有时呈红色至深红色，叶脉在上面略突起，明显。伞形花序呈平头状，具花 18~30 朵；花小，直径约 8mm，具浓香气；花冠向背面反卷呈球状，淡黄色至淡紫红色，边缘被短绢毛；副花冠淡黄色。

栽培习性：喜半阴、温暖潮湿的环境。成熟枝条扦插易生根，新萌发的插条需遮阴，待植株健壮后可适当加强光照以避免枝条徒长。本种叶片在昼夜温差较大且明亮处变为红色至深红色，极美观，健壮植株全年开花不断。

光照需求：

养护难度：

Distribution: Philippines.

Morphology: Climbing vine. Stem ca. 3mm diameter. Leaf blade leathery, wide lanceolate to ellipse, ca. 8cm long and 4-5cm wide, margin entire, apex acuminate to acute, green, sometimes red to deep red, vein convex and distinct. Umbel flat-topped, 18-30 flowered; flowersmall, ca. 8mm diameter, strong fragrance; corolla revolute in ball shape, light yellow to light fuchsia, margin sericeous; corona light yellow.

Habits: Grows well in half-shading, warm and moist environment; cuttings have high survival rate; need shading at early stage; additional of light retard excessive growth of branches; leaf become red to deep red at high temperature fluctuation between day and night; great ornamental value; mature plant blooms throughout the year.

Lighting requirement:

Maintenance difficulty:

Hoya obtusifolia Wight in Contr. Bot. India 38. 1834.

钝叶球兰

分布：缅甸、泰国和马来西亚。

形态特征：攀援藤本。茎直径约 7mm。叶片大，肉质，椭圆形至长圆形，长 10~15cm，宽 5~7cm，边缘全缘，先端急尖或钝，绿色，表面光滑，中脉明显，侧脉不显。伞形花序具花 10~15 朵；花直径约 2.5cm，具淡香气；花冠内弯呈爪状，白色直至浅黄色；副花冠白色，雄蕊红色。

栽培习性：喜明亮的散射光，喜高温高湿的环境。扦插枝条不易生根，适当提高环境温度有一定的促进作用，植株生长较慢，可提高温度及湿度以促进生长。植株茎粗且质硬，故需在其幼嫩时及时塑形，以获得美观的株形。本种花序较少，置于明亮处可促进萌生花序。

光照需求：
养护难度：

Distribution: Myanmar, Thailand and Malaysia.

Morphology: Climbing vine. Stem ca. 7mm diameter. Leaf blade large, fleshy, ellipse to oblong, 10-15cm long and 5-7cm wide, margin entire, apex acute or obtuse, green, surface smooth, midvein distinct, lateral vein obscure. Umbel 10-15 flowered; flower ca. 2.5cm diameter, light fragrance; corolla incurved, claw-like, white to light yellow; corona white; stamen red.

Habits: Plant likes bright scattered light, high temperature and high humidity environment; cuttings difficult to form roots; proper increase environmental temperature help to form roots; slowly-growing; addition of temperature and humidity help growing; stems thick and hard, shaping should be done to get beautiful plant shape at young stage; inflorescence of this *Hoya* is relatively fewer, bright light promote plant form more inflorescence.

Lighting requirement:
Maintenance difficulty:

Hoya odorata Schltr. in Philipp. J. Sci. 1(Suppl.): 303. 1906.

甜香球兰

分布：菲律宾。

形态特征：半灌木。茎木质化，直径约 3mm。叶片革质，披针形至椭圆形，长约 4cm，宽约 2cm，边缘全缘，先端渐尖，绿色，表面光滑，中脉明显，侧脉不明显。伞形花序疏松，具花 3~8 朵，花谢后花序梗一同脱落；花直径约 1.5cm，具清香气，夜间更甚。花冠内弯呈爪状，白色；副花冠黄绿色至淡绿色。

栽培习性：喜半阴、通风、湿润的环境，高温闷热则生长不佳。栽培需选用疏松、透气的基质，积水则影响根的发育。本种扦插枝条不易生根，成活率低。适当增加光照可促进植株开花。

光照需求：☀☀☀

养护难度：🌱🌱

Distribution: Philippines.

Morphology: Subshrub. Stem woody, ca. 3mm diameter. Leaf blade leathery, lanceolate to ellipse, ca. 4cm long and ca. 2cm wide, margin entire, apex acuminate, green, surface smooth, midvein distinct, lateral vein obscure. Umbel loose, 3-8 flowered; flower ca. 1.5cm diameter, fragrance, stronger at night; peduncle drops after flowers wither; corolla incurved, claw-like, white; corona yellowish green to light green.

Habits: Plant likes half shade, ventilate and moist environment; high temperature and stuffiness hinder growth; loose and ventilate substrate preferred; water logging hold up roots growth; cuttings are difficult to take roots; low survival rate; proper increment in light can promote the plant bloom.

Lighting requirement: ☀☀☀

Maintenance difficulty: 🌱🌱

Hoya ovalifolia Wight & Arn. in Wight in Contr. Bot. India 37. 1834.

卵叶球兰

分布: 中国(海南)。

形态特征: 攀援藤本。茎直径约3mm。叶片厚纸质,卵状披针形,长6~8cm,宽3~5cm,边缘全缘,先端急尖,叶脉清晰。伞形花序具花12~20朵;花直径约1cm,具清香气;花冠向背面反折,淡黄白色、浅黄色或粉红色;副花冠红色。

栽培习性: 适应性强,较易养护。喜半阴、凉爽湿润的环境。栽培基质忌积水,否则根及茎均易腐烂。扦插枝条易生根。本种在凉爽的环境下叶片生长较佳,叶形美观,但却不易开花,而在较高光照下叶片生长一般,但开花较频繁。

光照需求: ☀ ☀ ☀

养护难度: 🌱🌱

Distribution: China (Hainan).

Morphology: Climbing vine. Stem ca. 3mm diameter. Leaf blade thick papyraceous, ovate to lanceolate, 6-8cm long and 3-5cm wide, margin entire, apex acute, vein distinct. Umbel 12-20 flowered; flower ca. 1cm diameter, fragrance; corolla reflexed, creamy white, light yellow or pink; corona red.

Habits: Well-adapted and easy-cultivated; likes half shade, cool and moist environment; avoid water logging in substrate, otherwise roots and stems tend to rot; cuttings form roots easily; leaves grow well in coolness, but plant difficult to bloom; in high light source leaves sparse, but blooms a lot.

Lighting requirement: ☀ ☀ ☀

Maintenance difficulty: 🌱🌱

Hoya pachyclada Kerr in Bull. Misc. Inform. Kew, 1939: 462. 1939.

粗蔓球兰

分布：越南、老挝、缅甸和泰国。

形态特征：悬垂藤本。茎较粗，直径约8mm，节间短。叶柄粗壮，叶片肉质，椭圆形至阔椭圆形，长约10cm，宽约8cm，边缘全缘，先端急尖或钝，深绿色，中脉明显，侧脉明显或不明显。伞形花序呈球状，具花20~35朵；花直径约1.4cm，略具姜花香气；花冠向背面反卷或反折，白色或淡黄绿色；副花冠白色；雄蕊浅黄色。

栽培习性：适应性强，易养护。喜半阴、温暖、湿润且通风的环境，过于闷热则叶片易腐烂。能适应各种生长基质，但以利于根吸附的基质为佳。扦插枝条易生根。植株生长快，易开花，是理想的观赏种类。

光照需求：

养护难度：

Distribution: Vietnam, Laos, Myanmar and Thailand.

Morphology: Hanging vine. Stem thick, ca. 8mm diameter, internode short. Petiole stout; leaf blade fleshy, ellipse to broad ellipse, ca. 10cm long and ca. 8cm wide, margin entire, apex acute to obtuse, deep green, midvein distinct, lateral vein distinct or obscure. Umbel convex, 20-35 flowered; flower ca. 1.4cm diameter, ginger flower fragrance; corolla revolute or reflexed, white or light green; corona white; stamen light yellow.

Habits: Well-adapted and easy-cultivated; grows well in half-shading, warm and moist environment; stuffiness may cause leaves rotting; adapted to various substrate, and substrate can help roots for attachment preferred; cuttings form roots easily; fast-growing and blooms easily, having great ornamental value.

Lighting requirement:

Maintenance difficulty:

Hoya padangensis Schltr. in Beih. Bot. Centralbl. 34(2): 15. 1916.

巴东球兰

分布：马来西亚和印度尼西亚（苏门答腊岛和爪哇岛）。

形态特征：攀援藤本。茎直径约 2mm。叶片革质，披针形至长圆形，长约 9cm，宽 3~4cm，边缘略波状，先端渐尖，灰绿色，有时具银色斑点，表面粗糙，叶脉不明显。伞形花序具花 15~25 朵；花直径约 1.5cm，具清香气；花冠略内弯，深裂至近基部，裂片窄，黄白色，先端淡褐色至粉色；副花冠黄白色。

栽培习性：喜明亮的散射光，光照不足则生长较差。喜温暖、潮湿的环境，亦较耐旱。成熟枝条扦插成活率高。植株生长势适中，易塑形，健壮植株开花频繁。本种花形状特别，是独特的观赏种类。

备注：本种以其原产地印度尼西亚的巴东（Padang）命名。

光照需求：☀☀☀☀☀
养护难度：🌱🌱

Distribution: Malaysia and Indonesia (Sumatra Island and Java Island).

Morphology: Climbing vine. Stem ca. 2mm diameter. Leaf blade leathery, lanceolate to oblong, ca. 9cm long and 3-4cm wide, margin undulant, apex acuminate, greyish-green, sometimes silver spots, surface rough, vein obscure. Umbel 15-25 flowered; flower ca. 1.5cm diameter, fragrance; corolla slightly incurved, lobe narrow, split to the base, yellowish white, corolla apex brown to pink; corona yellowish white.

Habits: Plant likes bright scattered light; insufficient light hinder growth; grows well in warm and moist environment; drought-resistant; mature branch as cuttings have high survival rate; plant grows at a medium speed, easy to shape; healthy plant blooms frequently; unique flower endows great ornamental value.

Note: This *Hoya* comes from Padang, Indonesia.

Lighting requirement: ☀☀☀☀☀
Maintenance difficulty: 🌱🌱

Hoya pandurata Tsiang in Sunyatsenia, **4**(1–2): 125–126. 1939.

琴叶球兰

分布：中国（云南）。

形态特征：附生半灌木。茎直径约4mm。叶柄短，叶片厚纸质至革质，提琴形或长圆形，长约9cm，宽2~3cm，基部圆，边缘全缘或略呈波状，先端尾状，绿色，叶脉不明显。伞形花序具花5~10朵；花直径约1cm，具淡香气；花冠略向背面反折，黄色至橙黄色，表面被短茸毛；副花冠黄色；雄蕊红色。

栽培习性：喜半阴，不宜阳光直射，喜凉爽、湿润的环境，不甚耐寒，10℃以下茎易干枯，故需保温。栽培宜选用疏松、透气且利于根吸附的基质。成熟枝条扦插易成活，植株生长较慢。

备注：本种为我国特有，株形特别，叶片美观，是珍贵的观赏种类。

光照需求：☀ ☀ ☀

养护难度：🌱 🌱 🌱

Distribution: China (Yunnan).

Morphology: Subshrub. Stem ca. 4mm diameter. Petiole short, leaf blade thick papyraceous to leathery, pandurate or oblong, ca. 9cm long and 2-3cm wide, base round, margin entire or undulant, apex caudate, green, vein obscure. Umbel 5-10 flowered; flower ca. 1cm diameter, light fragrance; corolla sightly reflexed, yellow to orange, surface tomentose; corona yellow; stamen red.

Habits: Plant likes half shade; avoid direct sun light; cool and moist environment preferred; nonhardy, stems tends to wither when temperature drops below to 10 C ; loose and ventilate substrates help the roots for attachment; mature branch cuttings have high survival rate; slowly-growing.

Note: *Hoya pandurata* is endemic to China. Its unique plant shape and beautiful leaves make itself as valuable ornamental species.

Lighting requirement: ☀ ☀ ☀

Maintenance difficulty: 🌱 🌱 🌱

Hoya parasitica Wall. ex Wight in Contr. Bot. India 37, 1834.

寄生球兰

分布：缅甸、老挝和泰国。

形态特征：攀援藤本。茎直径约 5mm。叶片肉质，椭圆形，长约 15cm，宽 4~6cm，边缘全缘，先端渐尖至急尖，深绿色，表面光滑，叶脉清晰。伞形花序呈球状，具花 25~38 朵；花直径约 1.3cm，具浓香气；花冠略向背面反折，黄绿色至淡绿色；副花冠红色。

栽培习性：适应性强、易养护。喜阴、喜欢温暖潮湿的环境，耐寒且耐旱。栽培宜选用疏松、透气的基质。扦插枝条易生根，植株生长较快。枝条舒展，如自由攀援则萌生的花序较多，全年开花不断。本种养护容易，是室外栽培的理想种类。

光照需求：☀☀

养护难度：🌱

Distribution: Myanmar, Laos and Thailand.

Morphology: Climbing vine. Stem ca. 5mm diameter. Leaf blade fleshy, ellipse, 15cm long and 4-6cm wide, margin entire, apex acuminate to acute; deep green, surface smooth, veindistinct.Umbel 25-38 flowered; flower ca. 1.3cm diameter, strong fragrance; corolla slightly reflexed, yellowish green to light green; coronared.

Habits: Well-adapted and easy-cultivated; likes shade, warm and moist environment; cold-resistant and drought-resistant; loose and ventilate substrate preferred; cuttings are easy to form roots; fast-growing; branches spreading; freely-climbing plants form more inflorescence and bloom throughout the year; an excellent species for outdoor cultivation.

Lighting requirement: ☀☀

Maintenance difficulty: 🌱

Hoya patella Schltr. in Bot. Jahrb. Syst. **50**: 132. 1913.

碗花球兰

分布：巴布亚新几内亚。

形态特征：攀援藤本。茎直径约 2mm。叶片厚纸质至革质，长圆形，长约 7cm，宽约 4cm，边缘全缘，先端渐尖至急尖，深绿色，密被短茸毛，中脉明显，侧脉不明显。伞形花序疏松，具花 1~3 朵；花大，直径约 3.5cm，具淡香气；花冠阔钟状，内面光滑，边缘被短茸毛，淡粉色至淡紫红色，稀为白色；副花冠深红色至紫红色；雄蕊颜色与副花冠同。

栽培习性：喜半阴、温暖、潮湿的环境，不耐寒，环境温度保持在 20~28℃最佳。栽培宜选用疏松、透气的基质，扦插枝条较不易生根。本种对栽培环境要求较高，如环境适宜则生长较快且常年开花不断。

光照需求：☀ ☀ ☀
养护难度：🌱 🌱

Distribution: Papua New Guinea.

Morphology: Climbing vine. Stem ca. 2mm diameter. Leaf blade thick papyraceous to leathery, oblong, ca. 7cm long and ca. 4cm wide, margin entire, apex acuminate to acute; deep green, densely tomentose, midvein distinct, lateral vein obscure. Umbel loose, 1-3 flowered; flower ca. 3.5cm diameter, light fragrance; corolla wide campanulate, inner surface smooth, margin tomentose, light pink to light fuchsia, rare white; corona deep red to fuchsia; stamen has same color with corona.

Habits: Grows well in half-shading, warm and moist environment; nonhardy; temperature of 20-28℃ preferred; loose and ventilate substrate preferred; cuttings rarely form roots; plant has high requirements for cultivation; fast-growing and blooms through out the year in suitable environment.

Lighting requirement: ☀ ☀ ☀
Maintenance difficulty: 🌱 🌱

Hoya pallilimba Kleijn & Donkelaar in Blumea 46(3): 479. 2001.

豆瓣球兰

分布：印度尼西亚（苏拉威西岛）。

形态特征：攀援藤本。茎直径约2mm，表面被短茸毛。叶片革质，硬，卵形至阔椭圆形，长约5cm，宽3.5~4cm，边缘全缘，先端急尖，中间略隆起，表面有时具斑点，中脉明显，侧脉不明显。伞形花序呈平头状，疏松，具花18~25朵；花直径约7mm，具淡香气；花冠向背面反卷呈扁球状，粉红色至紫红色，表面具微茸毛；副花冠黄色。

栽培习性：喜半阴、湿润且通风的环境。栽培宜选用疏松、透气的基质。扦插枝条易成活，但根幼嫩时忌施肥，否则容易烂根，根发育成熟后可适当施肥促进植株生长。本种生长较快，但营养生长期长，需待植株健壮后才能萌生花序。

光照需求：

养护难度：

Distribution: Indonesia (Sulawesi Island).

Morphology: Climbing vine. Stem tomentose, ca. 2mm diameter. Leaf blade leathery, rigid, ovate to broad ellipse, ca. 5cm long and 3.5-4cm wide, margin entire, apex acute, leaf center convex, sometimes spots on the surface, midvein distinct, lateral vein obscure. Umbel flat-topped, loose, 18-25 flowered; flower ca. 7mm diameter, light fragrance; corolla revolute in ball shape, pink to fuchsia, tomentose; corona yellow.

Habits: Grows well in half-shading, ventilate and moist environment; loose and ventilate substrate preferred; cuttings have high survival rate; avoid fertilizers when roots are young, otherwise roots tend to rot; proper fertilizers help its growing after roots are maturation; fast-growing; needs a long vegetative period to blossom after plant grows maturely.

Lighting requirement:

Maintenance difficulty:

Hoya paziae Kloppenb. in Fraterna 1(3), Philipp. Hoya Sp. Suppl.: VI. 1990.

巴兹球兰

分布：菲律宾。

形态特征：半灌木。茎直径约 3mm。叶片厚纸质，椭圆形，长约 8cm，宽约 3cm，边缘全缘，先端尾状，深绿色，表面光滑，中脉明显，侧脉不显。伞形花序疏松，具花 8~15 朵；花直径约 2cm，具淡香气，夜晚更甚；花冠平展或略内弯，白色，表面光滑；副花冠红色至红棕色。

栽培习性：本种形态及习性与甜香球兰 Hoya odorata 较相似，但本种花谢后花序梗不脱落，而甜香球兰的花谢后花序梗与花一同脱落，此为两者区别之一；喜半阴、通风、湿润的环境，高温闷热则生长不佳。栽培宜选用保湿但不积水的基质。由于成熟枝条木质化，故扦插生根较慢。置于通风且明亮处有助于植株开花。

光照需求：

养护难度：

Distribution: Philippines.

Morphology: Subshrub. Stem ca. 3mm. Leaf blade thick papyraceous, ellipse, ca. 8cm long and ca. 3cm wide, margin entire, apex caudate, deep green, surface smooth, midvein distinct, lateral vein obscure. Umbel loose, 8-15 flowered; flower ca. 2cm diameter, light fragrance, stronger at night; corolla spreading or slightly incurved, white, surface smooth; corona red to red brown.

Habits: Morphology and habits of *Hoya paziae* are similar to *Hoya odorata*, but peduncles of the former one not drop after the flowers wither; likes half shade, ventilate and moist environment, and not grow healthily in high temperature and stuffiness; moist and non-waterlogging substrate preferred; mature branches lignified, branch cuttings grow slowly; ventilation and bright light help for blooming.

Lighting requirement:

Maintenance difficulty:

Hoya picta (Blume) Miq. in Fl. Ned. Ind. 2: 524. 1857.

彩芯球兰

分布：印度尼西亚（爪哇岛）。

形态特征：垂悬藤本。茎直径约1mm。叶片肉质，小，椭圆形，长约2cm，宽约1.5cm，边缘全缘，先端急尖，下面浅绿色，上面绿色，表面光滑。伞形花序呈平头状，具花10~20朵；花直径约8mm，几无香气；花冠向背面反卷呈球状，黄色，表面被短茸毛；副花冠黄色，基部具紫红色斑点。

栽培习性：喜半阴、通风的环境，如环境闷湿则枝条及叶片易腐烂。栽培宜选用疏松、透气的基质。本种枝条纤细，扦插需选用利于根部吸附的基质。稍干燥的环境有利于植株生长及开花。

光照需求：☀ ☀ ☀

养护难度：🌱 🌱 🌱

Distribution: Indonesia (Java Island).

Morphology: Hanging vine. Stem ca. 1mm diameter. Leaf blade fleshy, small, ellipse, ca. 2cm long and ca. 1.5cm wide, margin entire, apex acute, abaxially light green, adaxially green, smooth. Umbel flat-topped, 10-20 flowered; flower ca. 8mm diameter, nearly odourless; corolla revolute in ball shape, yellow, surface tomentose; corona yellow, with fuchsia spots at the base.

Habits: Plant likes half shade and ventilation; leaves and branches tend to rot in clammy environment; loose and ventilate substrate preferred; branches are soft and thin, substrate that used to plant cuttings should be help roots for attachment; slightly dry environment can help its growing and blooming.

Lighting requirement: ☀ ☀ ☀

Maintenance difficulty: 🌱 🌱 🌱

Hoya pottsii J. Traill in Trans. Hort. Soc. London **7**: 25–26, pl. 1. 1830.

铁草鞋 三脉球兰

分布：亚洲至大洋洲的热带、亚热带地区。

形态特征：攀援藤本。茎直径约 3mm。叶片肉质，卵状椭圆形至椭圆状披针形，长 6~12cm，宽 3~6cm，边缘全缘，先端急尖，深绿色，叶脉明显。伞形花序呈球状，具花 20~40 朵；花直径约 9mm，具浓香气；花冠向背面反折，白色、黄绿色、黄色或粉红色；副花冠通常白色。

栽培习性：适应性强，易养护。喜半阴、潮湿通风的环境，耐寒。对栽培基质要求不高。成熟枝条扦插成活率高，植株生长快。本种在明亮处栽培叶片变为橙红色，较美观。栽培环境适宜则终年开花不断，花香浓郁，是易养护且观赏性佳的种类。

备注：本种发现于我国珠海（斗门），以其标本采集人 John Potts 命名。

光照需求：

养护难度：

Distribution: Tropic and subtropic zones of Asia and Oceania.

Morphology: Climbing vine. Stem ca. 3mm diameter. Leaf blade fleshy, oval ellipse to ellipse lanceolate, 6-12cm long and 3- 6cm wide, margin entire, apex acute, deep green, vein distinct. Umbel convex, 20-40 flowered; flower ca. 9mm diameter, strong fragrance; corolla reflexed, white, yellowish green, yellow or pink; corona usually white.

Habits: Well-adapted and easy-cultivated; likes half shade, ventilation and moisture; cold-resistant; low demand in substrate; mature branch cuttings have high survival rate; fast-growing; leaf turns orange in bright place; blooms throughout the year in suitable environment; this *Hoya* has strong fragrance flower which makes it as a famous ornamental species.

Note: Found in Zhuhai City (Doumen) of China, and named after its specimens collector, John Potts.

Lighting requirement:

Maintenance difficulty:

Hoya praetorii Miq. Fl. Ned. Ind. 2: 526. 1857.

猴王球兰

分布： 马来西亚和印度尼西亚（苏门答腊岛和爪哇岛）。

形态特征： 半灌木，枝条披散。茎直径约 6mm。叶片纸质，椭圆形，长 12~15cm，宽 5~7cm，边缘全缘，先端略呈尾状，深绿色，表面光滑，叶脉明显。伞形花序呈平头状，具花 10~17 朵；花直径约 1.5cm，无香气。花冠向背面反折，黄色，中部以下被长绢毛，副花冠红色至紫红色，先端淡黄色。

栽培习性： 对环境要求较高。喜半阴、温暖、高湿且通风的环境，不耐寒，冬季需保温。栽培需选用疏松、透气、保湿但不积水且利于根吸附的基质。茎肉质，扦插需注意基质不能过于潮湿，否则茎容易腐烂。植株在遮阴、通风且温度为 20~30℃的环境下生长较好。茎及枝条披散，需适当扶持以获得美观的株形；本种花形奇特、美观，是著名的观赏种类。

备注： 本种以其标本采集人 Christiaan Frederik Eduard Praetorius 命名。

光照需求：
养护难度：

Distribution: Malaysia and Indonesia (Sumatra Island and Java Island).

Morphology: Subshrub, with scattered branches. Stem ca. 6mm diameter. Leaf blade papyraceous, ellipse, 12-15cm long and 5-7cm wide, margin entire, apex caudate, deep green, surface smooth, vein distinct. Umbel flat-topped, 10-17 flowered; flowerca. 1.5cm diameter, odourless; corolla reflexed, yellow, sericeous lower; corona red to fuchsia, apex light yellow.

Habits: Plant grows well in half-shading, warm, ventilate and humid environment; nonhardy, needs insulation in winter; loose, ventilate, water-preserving, non-water-logging substrate can help the roots for attachment preferred; substrate not be too moist, otherwise stems tend to rot; plant grows well in shading and ventilate environment, kept temperature between 20-30 ℃ ; plant has scattered branches, needs proper support for beautiful plant shape; unique and beautiful flowers endow itself as great ornamental value.

Note: Named after the collector, Christiaan Frederik Eduard Praetorius.

Lighting requirement:
Maintenance difficulty:

Hoya pseudolittoralis C. Norman in Brittonia 2: 328. 1937.

伪滨海球兰

分布：巴布亚新几内亚。

形态特征：攀援藤本。茎直径约 5mm。叶片肉质，阔披针形至椭圆形，长约 10cm，宽约 5cm，边缘全缘，先端尾尖，灰绿色或橙红色，表面光滑，叶脉不显。伞形花序呈平头状，具花 10~20 朵；花直径约 1.2cm，具玫瑰香气；花冠平展，白色，被短茸毛；副花冠中间红色至粉红色，边缘粉红色至白色。

栽培习性：喜明亮的散射光，喜温暖潮湿的环境，不耐寒，冬季需适当保温，否则易受寒害。栽培宜选用疏松、透气的基质，成熟枝条扦插易生根。植株生长较缓慢，本种叶片在高光照及昼夜温差较大的环境下变为橙红色，较美观。

光照需求：

养护难度：

Distribution: Papua New Guinea.

Morphology: Climbing vine. Stem ca. 5mm diameter. Leaf blade fleshy, broad lanceolate and ellipse, ca. 10cm long and ca. 5cm wide, margin entire, apex caudate; grayish green or orange, surface smooth, vein obscure. Umbel flat-topped, 10-20 flowered; flower ca. 1.2cm diameter, rose fragrance; corolla spreading, white, tomentose; corona center red to pink, margin pink to white.

Habits: Plant likes bright scattered light, warm and moist environment; nonhardy, needs proper insulation in winter, otherwise tends to suffer from chilling injury; loose and ventilate substrate preferred; mature branch cuttings form roots easily; slowly-growing; leaf blade turns to orange in strong light or inlarge temperature fluctuation between day and night.

Lighting requirement:

Maintenance difficulty:

Hoya pubicalyx Merr. in Philipp. J. Sci. 13: 331. 1918.

毛萼球兰（新拟）

分布：菲律宾。

形态特征：攀援藤本。茎直径约 6mm。叶片肉质，披针形至倒披针形，长约 15cm，宽 4~5cm，边缘全缘，先端渐尖，深绿色，表面光滑，中脉明显，侧脉不明显。伞形花序呈球状，具花 22~35 朵；花直径约 1.5cm，具淡香气；花冠平展，紫红色，被短茸毛；副花冠白色；雄蕊紫红色。

栽培习性：适应性强，易养护。喜阴、温暖、湿润的环境，耐寒且耐旱。对栽培基质要求不高，扦插枝条易生根，植株生长快，枝条舒展。本种全年开花不断，如任其自由攀援则开花更多，是室外栽培的理想种类。

光照需求：☀☀
养护难度：🔧

Distribution: Philippines.

Morphology: Climbing vine. Stem ca. 6mm diameter. Leaf blade fleshy, lanceolate to oblanceolate, ca. 15cm long and 4-5cm wide, margin entire, apex acuminate, deep green, smooth, midvein distinct, lateral vein obscure. Umbel convex, 22-35 flowered; flower ca. 1.5cm diameter, light fragrance; corolla spreading, fuchsia, tomentose; corona white; stamen fuchsia.

Habits: Well-adapted and easy to care; likes shade, warm and moist environment; cold-resistant and drought-resistant; low demand in environment; cuttings form roots easily; fast-growing, with spreading branches; blooms throughout the year; freely-climbing plant can produce more flowers; a good outdoor-cultivating species.

Lighting requirement: ☀☀
Maintenance difficulty: 🔧

Hoya retusa Dalzell in Kew Jour. Bot. **4**: 294. 1852.

截叶球兰（新拟） 断叶球兰

分布：印度北部。

形态特征：悬垂藤本。茎细弱，直径约1mm。叶片肉质，条形，长约12cm，宽约5mm，边缘全缘或略呈波状，先端截形，微凹，绿色，表面疏被短茸毛，叶脉不显。伞形花序疏松，具花1~3朵，花谢后花序梗一同脱落；花直径约1.6cm，具淡香气；花冠平展，白色，被短茸毛；副花冠紫红色；雄蕊白色。

栽培习性：喜半阴、湿润且通风的环境。栽培宜选用疏松、保湿且利于根吸附的基质。茎细弱，扦插易干枯，需保湿以提高成活率。植株生长较慢，增加环境湿度可促进生长，花期一般为9~11月。本种叶形独特，花相对叶片显得较大，是优良的悬垂观赏种类。

光照需求：

养护难度：

Distribution: Northern India.

Morphology: Hanging vine. Stem thin, ca. 1mm diameter. Leaf blade fleshy, linear, ca. 12cm long and ca. 5mm wide, margin entire or undulant, apex truncate, retuse, green, sparsely tomentose, vein obscure. Umbel loose, 1-3 flowered; flower ca. 1.6cm diameter, light fragrance; corolla spreading, white, tomentose; corona fuchsia; stamen white.

Habits: Grows well in half-shading, ventilate and moist environment; loose and water-preserving substrate help for plant attachment; thin stem, cuttings tend to dry up, proper moisture preservation help to increase survival rate; slowly-growing; increase humidity help growing; blooms in September, October and November; unique leaf shape, flowers are relatively larger than the leaves; these two characters make itself as a special ornamental species.

Lighting requirement:

Maintenance difficulty:

Hoya revoluta Wight ex J. D. Hook. Fl. Brit. India **4**(10): 55. 1883.

反卷球兰

分布：马来西亚。

形态特征：攀援藤本。茎直径约 2mm。叶片肉质，披针形至狭椭圆形，常向背面弯弓，长约 5cm，宽约 2.5cm，边缘全缘，先端渐尖，绿色至深绿色，叶脉不明显；伞形花序呈平头状，具花 15~25 朵；花直径约 6mm，具淡香气；花冠初时平展，后向背面反卷，被短绢毛；副花冠黄色；雄蕊红色。

栽培习性：喜明亮的散射光，喜温暖、湿润的环境。栽培宜选用疏松且颗粒较小的基质。扦插枝条易生根，成活率高。植株生长快，需及时整形以保持株形美观。本种在明亮的环境下叶片坚挺，长势更好。

光照需求：

养护难度：

Distribution: Malaysia.

Morphology: Climbing vine. Stem ca. 2mm diameter. Leaf blade fleshy, lanceolate to narrow ellipse, usually recurved, ca. 5cm long and ca. 2.5cm wide, margin entire, apex acuminate, green to deep green, vein obscure. Umbel flat-topped, 15-25 flowered; flower ca. 6mm diameter, light fragrance; corolla spreading at early stage, then revolute, surface sericeous; corona yellow; stamen red.

Habits: Plant likes bright scattered light, grows well in warm and moist environment; loose and small particle substrate preferred; cuttings form roots easily and have high survival rate; fast-growing, needs regular pruning for beautiful plant shape; leaves are rigid and grow well in bright space.

Lighting requirement:

Maintenance difficulty:

Hoya rigida Kerr in Bull. Misc. Inform. Kew **1939**: 463. 1939.

硬叶球兰

分布：泰国。

形态特征：攀援藤本。茎直径约 5mm。叶片厚革质至肉质，阔披针形，长约 15cm，宽约 7cm，基部宽，边缘全缘，先端渐尖，绿色至深绿色，表面光滑，叶脉明显。伞形花序呈球状，具花 20~35 朵；花直径约 2cm，具浓香气；花冠向背面反折或略反卷，淡黄绿色；副花冠白色，雄蕊红色。

栽培习性：喜半阴、温暖、湿润的环境。对栽培基质要求不高，疏松、透气的基质有利于根部发育。株形较大，需要预留充足空间供其生长，本种叶片在光照不足时形态变化较大。易开花，是优良的观赏种类。

光照需求：● ● ●
养护难度：🪴🪴

Distribution: Thailand.

Morphology: Climbing vine. Stem ca. 5mm diameter. Leaf blade thick leathery to fleshy, broad lanceolate, ca. 15cm long and ca. 7cm wide, leaf base broad; margin entire, apex acuminate, green to deep green, surface smooth, vein distinct. Umbel convex, 20-35 flowered; flower ca. 2cm diameter, strong fragrance; corolla reflexed or slightly revolute, light creamy green; corona white; stamen red.

Habits: Grows well in half-shading, warm and moist environment; low demand in substrate; loose and ventilate substrate can help for root growth; due to large size of plant, large space should be given; leaf blade changes a lot in insufficient light; blooms easily and having great ornamental value.

Lighting requirement: ● ● ●
Maintenance difficulty: 🪴🪴

Hoya scortechinii King & Gamble in J. Asiat. Soc. Bengal, Pt. 2, Nat. Hist. 74(2): 567. 1908.

苏格球兰

分布：马来西亚。

形态特征：攀援藤本。茎直径约 2mm。叶片肉质，披针形至倒披针形，长约 6cm，宽 2~3cm，边缘全缘，先端渐尖，绿色，常具银色斑点，叶脉不显。伞形花序呈球状，具花 8~15 朵；花直径约 8mm，具淡香气；花冠向背面反卷，粉红色；副花冠白色；雄蕊红色。

栽培习性：适应性较强，喜阴、喜温暖、潮湿的环境，较耐寒，可通过控水越冬。对栽培基质要求不高。植株生长快。本种茎较柔软细弱，需适当扶持保持美观株形，营养生长期较长，需待植株健壮方会萌生花序。

备注：本种以其标本采集人意大利神父及博物学家 Benedetto Scortechini 命名。

光照需求：

养护难度：

Distribution: Malaysia.

Morphology: Climbing vine. Stem ca. 2mm diameter. Leaf blade leshy, lanceolate to oblanceolate, ca. 6cm long and 2-3cm wide, margin entire, apex acuminate, green, usually with silver spots, vein obscure. Umbel convex, 8-15 flowered; flower ca. 8mm diameter, light fragrance; corolla revolute, pink; corona white; stamen red.

Habits: Well-adapted and likes shade, warm and moist environment; cold-resistant, plant can live throughout the cold seasons by water controlling; low demand in substrate; fast-growing; due to the weak stems, proper support needed for beautiful plant shape; needs a long vegetative period to blossom after plant grows maturation.

Note: Italian priest and naturalist, Benedetto Scortechini collected this *Hoya* and it was named after him.

Lighting requirement:

Maintenance difficulty:

Hoya shepherdi Short ex Hooker in Bot. Mag. 5269. 1861.

柳叶球兰（新拟）谢普德球兰

分布：中国（云南）、印度、尼泊尔和缅甸。

形态特征：攀援藤本。茎直径约 3mm。叶片肉质，披针形至倒披针形，两边沿中脉向内折呈"V"形，长约15cm，宽1~2cm，边缘全缘，先端渐尖，绿色，表面常具小凹点，叶脉不显。伞形花序具花5~10朵，花直径约1cm，具浓香气。花冠内弯呈爪状，白色，被短茸毛；副花冠白色；雄蕊粉红色。

栽培习性：喜阴、喜湿润、通风且凉爽的环境，闷热则花蕾易脱落。栽培宜选择疏松、透气的基质，扦插枝条成活率高，植株生长快，茎柔软，需及时扶持以避免株形披散。本种健壮植株花序多，花期甚为美观。

光照需求：

养护难度：

Distribution: China (Yunnan), India, Nepal and Myanmar.

Morphology: Climbing vine. Stem ca. 3mm diameter. Leaf blade fleshy, lanceolate to oblanceolate, induplicate and appearing as V shape, ca. 15cm long and 1-2cm wide, margin entire, apex acuminate, green, often with small pits on the surface, vein obscure. Umbel 5-10 flowered; flower ca. 1cm diameter, strong fragrance; corolla recurved, claw-like, white, tomentose; corona white; stamen pink.

Habits: Grows well in shading, ventilate, cool and moist environment; flower buds tend to drop in stuffiness; loose and ventilate substrate preferred; cuttings have high survival rate; fast-growing; branches usually scattered, needs proper support; healthy plant produce plenty inflorescence and having great ornamental value.

Lighting requirement:

Maintenance difficulty:

Hoya siariae Kloppenb. in Fraterna 15(3): 1. 2002.

锡亚球兰（新拟）希瑞球兰

分布：菲律宾（吕宋岛）。

形态特征：攀援藤本。茎直径约4mm。叶片革质，椭圆形至狭椭圆形，长约12cm，宽4~6cm，边缘全缘，先端渐尖至急尖，浅绿色，表面光滑，叶脉明显。伞形花序呈平头状，具花5~10朵；花直径约1.5cm，具淡香气；花冠钟状，边缘向背面反卷，淡紫红色或红色；副花冠紫红色。

栽培习性：适应性强，易养护。喜阴、温暖、湿润的环境，亦较耐寒。栽培宜选用疏松、透气且保湿的基质，扦插枝条成活率高，植株生长快。本种花序萌发早，易开花，是花形独特且易栽培的种类。

备注：本种以菲律宾植物育种学家Simeona V. Siar博士命名。

光照需求：☀☀

养护难度：🌱

Distribution: Philippines (Luzon Island).

Morphology: Climbing vine. Stem ca. 4mm diameter. Leaf blade leathery, ellipse to narrow ellipse, ca. 12cm long and 4-6cm wide, margin entire, apex acuminate to acute, light green, surface smooth, vein distinct. Umbel flat-topped, 5-10 flowered; flower ca. 1.5cm diameter, light fragrance; corolla campanulate, apex revolute, light purple-red or red; corona purple-red.

Habits: Well-adapted and easy-cultivated; grows well in shading, warm and moist environment; cold-resistant; loose, ventilate and water-preserving substrate preferred; cuttings have high survival rate; fast-growing; form inflorescence early; unique flower shape and easy to cultivate.

Note: Named after the Philippines botanical breeding scientist, Dr. Simeona V. Siar.

Lighting requirement: ☀☀

Maintenance difficulty: 🌱

Hoya sigillatis T. Green in Fraterna **17**(3): 2. 2004.

斑印球兰

分布：马来西亚（婆罗洲岛）。

形态特征：悬垂藤本。茎直径约 2mm。叶片革质，披针形，长 8~10cm，宽 1.5~2cm，边缘全缘，先端渐尖，紫红色或浅绿色，具银色斑点，叶脉不显。伞形花序呈平头状，具花 12~20 朵；花直径约 9mm，具微酸气；花冠平展，边缘向背面反卷，或全部向背面反卷，黄色至橙黄色；副花冠黄色至浅黄色。

栽培习性：喜半阴、温暖、潮湿的环境。栽培宜选用疏松、保湿但不积水的基质，枝条较纤细，扦插易干枯，故需保持高湿度。植株在高温、高湿的环境下生长快，在较高光照下叶片变为紫红色，银色斑点更突出，甚美观。本种花形似梅花，是独特的观赏种类。

光照需求：
养护难度：

Distribution: Malaysia (Borneo Island).

Morphology: Hanging vine. Stem ca. 2mm diameter. Leaf blade leathery, lanceolate, 8-10cm long and 1.5-2cm wide, margin entire, apex acuminate, purple-red or light green, with silver spots, vein obscure. Umbel 12-20 flowered; flower ca. 9mm diameter, sourish odour; corolla spreading, margin revolute or corolla entirely revolute, yellow to orange; corona yellow to light yellow.

Habits: Grows well in half-shading, warm and moist environment; loose, moist and non-water-logging substrate preferred; branches slender, cuttings tend to dry up, high humidity necessary; fast-growing in warm and moist environment; leaf blade becomes purple-red in high light and silver spots are more distinct; flowers like plum flowers makes it as a special ornamental species.

Lighting requirement:
Maintenance difficulty:

Hoya spartioides (Benth.) Kloppenb. in Fraterna **14**(2): 8. 2001.

花葶球兰（新拟）棒叶球兰

分布：马来西亚（婆罗洲岛）。

形态特征：半灌木。茎直径约 3mm，节间短。叶退化，仅幼株偶见 2 枚，卵形，长约 2cm，宽约 1cm，浅绿色，早落，成株无叶片；伞形花序多数，具花 4~8 朵；花序梗圆柱形，长约 20cm，直径约 2mm，绿色，代替叶片进行光合作用；花直径约 8mm，几无香气；花冠平展，橙黄色；副花冠白色；雄蕊黄色。

栽培习性：对环境要求高，养护不易。喜阴、温暖、湿润且通风的环境。栽培需选择疏松、透气、保湿但不积水的基质。扦插枝条生根慢，插条容易干枯，可选用疏松、透气的基质包裹插条，并增加环境湿度防止失水，适当增加扦插基质的温度及遮阴可促进插条生根，植株开花不易且花期极短。本种形态奇特，全株密生花序，加之养护不易，是珍稀的观赏种类。

光照需求：

养护难度：

Distribution: Malaysia (Borneo Island).

Morphology: Subshrub. Stem ca. 3mm diameter, internode short. Leaf reduced, young plant occasionally has two leaves, ovate, ca. 13cm long and ca. 9cm wide, light green, caducous; mature plant without leaves. Umbel 4-8 flowered; peduncle cylindrical, ca. 20cm long and ca. 2mm in diameter, green, replace leaf to realize photosynthesis; flower ca. 8mm diameter, nearly odourless; corolla spreading, orange; corona white; stamen yellow.

Habits: Difficult to grow, and has high requirements for environment. grows well in half-shading, warm, moist and ventilate environment; loose, ventilate, water-preserving and non-water-logging substrate preferred; cuttings are slowly-growing and easy to wither; loose and ventilate substrate to wrap cuttings and gradually increase humidity and temperature of substrates help form roots; rarely bloom and short flowering period; dense inflorescence and unique appearance make it as a rare ornamental species.

Lighting requirement:

Maintenance difficulty:

Hoya subcalva Burkill in Bull. Misc. Inform. Kew 1901: 141. 1901.

苏卡瓦球兰（新拟） 苏卡尔球兰

分布：巴布亚新几内亚。

形态特征：攀援藤本。茎直径约 5mm。叶片厚纸质至革质，长圆形，长约 12cm，宽约 6cm，边缘全缘，先端急尖或钝，深绿色，叶脉明显。伞形花序疏松，具花 10~18 朵；花直径约 2cm，具淡玫瑰香气；花冠略内弯，黄色，中间具粉红色斑点，被短茸毛；副花冠黄色；雄蕊红色。

栽培习性：适应性强，易养护。喜半阴，适当遮阴能使叶片增大而更美观。对栽培环境要求不高，具有一定的耐寒及耐旱能力，选择稍木质化的枝条扦插成活率高，增加施肥可促进植株快速生长，本种几乎全年开花不断。

光照需求：☀ ☀ ☀

养护难度：🌱

Distribution: Papua New Guinea.

Morphology: Climbing vine. Stem ca. 5mm diameter. Leaf blade thick papyraceous to leathery, ovale, ca. 12cm long and ca. 6cm wide, margin entire, apex acute or obtuse, deep green, vein distinct. Umbel loose, 10-28 flowered; flower ca. 2cm diameter, rose fragrance; corolla slightly incurved, yellow, center with pink spots, tomentose; corona yellow; stamen red.

Habits: Well-adapted and easy to cultivate; likes half shade; leaf blade grows larger in shade place and makes the plant more beautiful; low demand in environment; both cold and drought resistant; slightly woody branch cuttings have high survival rate; addition of fertilizers help to grow faster; plant blooms almost throughout the year.

Lighting requirement: ☀ ☀ ☀

Maintenance difficulty: 🌱

Hoya thomsonii J. D. Hook. Fl. Brit. India 4(10): 61. 1883.

西藏球兰

分布：中国、印度、缅甸和泰国（北部）。

形态特征：攀援藤本。茎直径约 2mm。叶片略肉质，长圆形至倒卵状长圆形，长 5~8cm，宽 2~4cm，边缘全缘，先端具小尖头，深绿色，有时具银色斑点，表面密被短茸毛，叶脉不明显。伞形花序具花 10~20 朵；花直径 1.5~2cm，具清香气；花冠内弯呈爪状，白色，边缘及内面被茸毛；副花冠白色；雄蕊淡黄白色。

栽培习性：喜阴、通风、湿凉的环境，忌闷热，否则根茎易腐烂。栽培宜选用疏松、透气的基质，忌积水。扦插需注意基质不宜过于潮湿，否则插条易腐烂。成株适应性较强，容易养护，置于凉爽且昼夜温差较大处可促进植株开花。

备注：本种以英国植物学家 Thomas Thomson 命名。

光照需求：

养护难度：

Distribution: China, India, Myanmar and Thailand (Northern region).

Morphology: Climbing vine. Stem ca. 2mm diameter. Leaf blade fleshy, ovale to obovate, 5-8cm long and 2-4cm wide, margin entire, apex mucronulate, deep green, sometimes silver spots on the surface, densely tomentose, vein obscure. Umbel 10-20 flowered; flower 1.5-2cm diameter, fragrance; corolla incurved, claw-like, white, margin and inner surface tomentose; corona white; stamen light yellowish white.

Habits: Grows well in shading, ventilate, moist and cool environment; avoid stuffiness, otherwise stem and root tend to rot; loose and ventilate substrate preferred; avoid water logging; cuttings tend to rot in wet substrate; mature plant well-adapted and easy-cultivate; tend to blooms in cool place with large temperature fluctuation between day and night.

Note: Named after the British botanist, Thomas Thomson.

Lighting requirement:

Maintenance difficulty:

Hoya tsangii C. M. Burton in Hoyan 9(4: 2):1. 1988.

怀德球兰（新拟） 澉球兰

分布：菲律宾。

形态特征：悬垂藤本。茎直径约2mm。叶片肉质，长椭圆形，长约5cm，宽1~2cm，边缘全缘，先端渐尖或钝，下面浅绿色，上面深绿色，表面被短茸毛，叶脉不明显。伞形花序呈平头状，具花10~16朵；花直径约5mm，具淡香气；花冠向背面反卷呈球状，紫红色，被短茸毛；副花冠中间紫红色，边缘黄色。

栽培习性：喜半阴、湿润且通风的环境，忌闷热，否则叶片易腐烂。栽培宜选用疏松、透气的基质，忌积水。扦插枝条易生根，植株生长较快，侧枝多，株形丰满美观，健壮植株全年开花不断。

备注：本种以我国植物采集人曾怀德（Tsang Wai Tak）命名。

光照需求：● ● ●

养护难度：🌱🌱

Distribution: Philippines.

Morphology: Hanging vine. Stem ca. 2mm diameter. Leaf blade fleshy, oblong, ca. 5cm long and 1-2cm wide, apex acuminate or obtuse, abaxially light green, adaxially deep green, tomentose, vein obscure. Umbel flat-topped, 10-16 flowered; flower ca. 5mm diameter, fragrance; corolla revolute in ball shape, purple-red, tomentose; corona center fuchsia, margin yellow.

Habits: Grows well in half-shading, moist and ventilate environment; avoid stuffiness, otherwise leaves tend to rot; loose and ventilate substrate preferred; avoid water logging; cuttings form roots easily; fast-growing and produce many lateral branches; the plant has beautiful appearance; healthy plant blooms throughout the year.

Note: Plant was collected by Chinese collector Tsang Wai Tak.

Lighting requirement: ● ● ●

Maintenance difficulty: 🌱🌱

Hoya vitellinoides Bakh. f. in Blumea **6**: 381. 1950.

黄结球兰

分布：马来西亚。

形态特征：攀援藤本。茎直径约 3mm。叶片厚纸质，椭圆形，长约 20cm，宽约 10cm，边缘全缘或略呈波状，先端渐尖至急尖，浅绿色，叶脉深绿色，明显。伞形花序呈球状，具花 25~35 朵；花直径约 8mm，具淡香气；花冠向背面反折，黄绿色；副花冠白色。

栽培习性：喜明亮的散射光，喜欢温暖、湿润的环境。根系生长快，栽培宜选用疏松、透气且颗粒较大的基质以促进根系的发展。扦插枝条易生根，株形较大，枝条长，需预留充足空间供其生长。本种叶片在明亮处脉络更加清晰美观，是优良的观叶种类。

光照需求：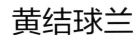

养护难度：🌱🌱

Distribution: Malaysia.

Morphology: Climbing vine. Stem ca. 3mm diameter. Leaf blade thick papyraceous, ellipse, ca. 20cm long and ca. 10cm wide, margin entire or slightly undulant, apex acuminate to acute, light green, vein deep gren, distinct. Umbel conves, 25-35 flowered; flower ca. 8mm diameter, light fragrance; corolla reflexed, yellowish green; corona white.

Habits: Plant likes bright scattered light, and grows well in warm and moist environment; roots grow fast; loose, ventilate and big-sized substrate can help the roots development; cuttings form roots easily; plant is big and branches are long, enough space should be given; more distinct and beautiful veins appear on leaf in bright space.

Lighting requirement:

Maintenance difficulty:

Hoya wallichii (Wight) C. M. Burton in Hoyan **18**(1:2): 5. 1996.

瓦氏球兰（新拟）沃里克球兰

分布： 马来西亚和印度尼西亚。

形态特征： 攀援灌木。茎直径约 3mm。叶面纸质，硬，长圆形，长约 9cm，宽 4~5cm，边缘全缘，先端尾尖，深绿色，叶脉明显。伞形花序具花 8~12 朵；花大，直径约 2cm，具姜花香气；花冠钟状，边缘具缺刻，白色；副花冠淡黄白色。

栽培习性： 对环境要求较高，喜阴、湿润、通风的环境。栽培宜选用疏松、透气的基质。扦插基质亦需疏松且透气，提高环境湿度可促进插条生根。植株开花早，根系生长稳定后即萌生花序。本种对给水要求较高，保持栽培基质干湿循环可促进植株生长，不规律地给水易导致花蕾脱落。

备注： 本种以丹麦植物学家 Nathaniel Wallich 命名。

光照需求： ☀☀

养护难度： 🌱🌱🌱

Distribution: Malaysia and Indonesia.

Morphology: Climbing shrub. Stem ca. 3mm diameter. Leaf blade papyraceous, rigid, ovale, ca. 9cm long and 4-5cm wide, margin entire, apex caudate, deep green, vein distinct. Umbel, 8-12 flowered; flower large, ca. 2cm diameter, garland-flower fragrance; corolla campanulate, margin incised, white; corona creamy white.

Habits: High requirements for environment; likes shade, moist and ventilate environment; loose and ventilate substrate preferred; addition of environmental humidity can help cuttings toform roots; plant blooms after roots grow maturely; high demand in water supply, drying and watering cycle can help growing; irregular irrigtion may lead to the flower buds abscission.

Note: Named after the Danish botanist, Nathaniel Wallich.

Lighting requirement: ☀☀

Maintenance difficulty: 🌱🌱🌱

Hoya walliniana Kloppenb. & Nyhuus in Fraterna 16(4): 9. 2003.

瓦林球兰

分布： 马来西亚（婆罗洲岛）。

形态特征： 攀援藤本。茎直径约2mm。叶片革质，阔披针形至椭圆形，长约3cm，宽约2cm，边缘全缘，先端渐尖，绿色，略具银色斑点，表面光亮，叶脉突起，明显。伞形花序呈平头状，具花20~35朵；花直径约7mm，几无香气；花冠向背面反卷呈扁球形，白色，被短茸毛；副花冠红色，先端白色。

栽培习性： 喜半阴、温暖、湿润的环境，不耐寒，冬季需保温，否则植株易干枯。栽培宜选用疏松、透气的基质，忌积水，否则根易腐烂。扦插枝条置于温暖的环境下易成活，环境适宜则植株生长快。本种置于较高光照下叶片变红，观赏性高。

备注： 本种以瑞典摄影师Pierre Wallin命名，以纪念其首次发现并拍摄本种的照片。

光照需求：

养护难度：

Distribution: Malaysia (Borneo Island).

Morphology: Climbing vine. Stem ca. 2mm diameter. Leaf blade leathery, wide lanceolate to ellipse, ca. 3cm long and ca. 2cm wide, margin entire, apex acuminate, green, surface shiny, slight silver spots, vein convex, distinct. Umbel flat-topped, 20-35 flowered; flower ca. 7mm diameter, nearly odourless; corolla revolute in ball shape, white, tomentose; corona red, apex white.

Habits: Grows well in half-shading, warm and moist environment; nonhardy, needs keep warm in winter, otherwise plant tends to wither; loose and ventilate substrate preferred; roots tend to rot in water logging; cuttings have high survival rate in warm environment; fast-growing in suitable environment; leaf blade turns red in brighter light.

Note: This *Hoya* was named after the Swedish photographer Pierre Wallin who found it and took its photograph.

Lighting requirement:

Maintenance difficulty:

Hoya wayetii Kloppenb. in Fraterna **1993**(2): 10. 1993.

威特球兰

分布：菲律宾（吕宋岛）。

形态特征：攀援藤本。茎柔软，常下垂，直径约 2mm。叶片肉质，披针形或倒披针形，两边沿中脉向内折呈"V"形，长约 8cm，宽 2~3cm，边缘全缘，黑褐色，先端渐尖，深绿色，叶脉不明显。伞形花序呈平头状，具花 18~30 朵；花直径约 8mm，几乎无香气；花冠向背面反卷呈球状，暗红色，被绢毛；副花冠紫红色。

栽培习性：喜明亮的散射光，喜湿润、通风的环境，忌闷热。扦插枝条需较长时间才能生根，栽培宜选择透气、保湿但不积水的基质，给予明亮光照则植株生长健壮且开花频繁。植株茎较柔软，虽为攀援藤本但常作悬垂种类栽培，内折且具黑褐色边缘的叶片是本种的特色之一。

备注：本种以其采集人 Maximo Wayet 命名。

光照需求：

养护难度：

Distribution: Philippines (Luzon Island).

Morphology: Climbing vine. Stem soft, usually pendulous, ca. 2mm diameter. Leaf blade fleshy, lanceolate or oblanceolate, induplicate and appear as V shape, ca. 8cm long and 2-3cm wide, margin entire, apex acuminate, deep green, margin dark brown, vein obscure. Umbel flat-topped, 18-30 flowered; flower ca. 8mm diameter, nearly odourless; corolla revolute in ball shape, dark red, sericeous; corona purple-red.

Habits: Plant likes bright scattered light; grows well in moist and ventilate environment; avoid stuffiness; cuttings need a long time to form roots; ventilate, water-preserving and non-water-logging substrate preferred; plant grow healthily and blooms frequently in bright light; soft stems, usually cultivated as hanging species; dark brown in folding leaf margin is one of its features.

Note: Type specimen of this *Hoya* was collected by the collector Maximo Wayet.

Lighting requirement:

Maintenance difficulty:

Hoya waymaniae Kloppenb. in Fraterna 1995(2): 8. 1995.

薇蔓球兰（新拟） 威蔓尼球兰

分布：马来西亚（婆罗洲岛）。

形态特征：攀援藤本，茎直径约 4mm。叶片革质，卵形至长圆形，长 7~10cm，宽 5~6cm，边缘波状并具缺刻，先端钝，深绿色，常具银色斑点，中脉明显，侧脉明显或不明显。伞形花序呈平头状，具花 12~20 朵，花序梗长，可达 40cm；花直径约 9mm，具柑橘香气；花冠向背面反卷呈球形，橙黄色，被短茸毛；副花冠中间紫红色，边缘淡黄色至白色。

栽培习性：喜明亮的散射光，喜温暖、通风的环境，忌闷湿。栽培宜选用疏松、透气的基质，忌积水，保持基质干湿循环可促进根生长。扦插枝条易生根，植株生长较慢，叶片在明亮处变为橙红色，较美观。本种花序梗极长，为其显著的特征之一。

备注：本种以球兰属植物园艺学家 Ann Wayman 命名。

光照需求：☀☀☀☀☀

养护难度：🌱🌱

Distribution: Malaysia (Borneo Island).

Morphology: Climbing vine. Stem ca. 4mm diameter. Leaf blade leathery, ovate to long elliptic, 7-10cm long and 5-6cm wide, margin undulant and incised, apex obtuse, deep green, often with silver spots, midvein distinct, lateral vein distinct or obscure. Umbel flat-topped, 12-20 flowered, peduncle to 40cm long; flower ca. 9mm diameter, citrus fragrance; corolla revolute in ball shape, orange, tomentose; corona center purple-red, margin creamy to white.

Habits: Plant likes bright scattered light; grows well in warm and ventilate environment; avoid stuffiness; loose and ventilate substrate is preferred; avoid water logging; alternate drying and watering cycle can help for roots growth; cuttings form roots easily; slowly-growing; leaf blade turns orange in bright place; long peduncle is one of its feature.

Note: Named after the *Hoya* horticulturalist Ann Waman.

Lighting requirement: ☀☀☀☀☀

Maintenance difficulty: 🌱🌱

中文索引

	A				
120	阿尔孔球兰	078	戴克球兰		H
022	阿丽亚娜球兰	074	戴维德球兰	118	荷秋藤
022	艾雷尔球兰	068	丹侬球兰	088	红副球兰
090	凹副球兰	098	淡黄球兰	132	红花帝王球兰
154	凹叶球兰	040	淡味球兰	130	红花基心覆叶球兰
	B	212	倒卵叶球兰	192	红花球兰
224	巴东球兰	078	德克球兰	240	猴王球兰
234	巴兹球兰	132	帝王球兰	138	厚冠球兰
078	芭蕉扇球兰	232	豆瓣球兰	140	厚冠球兰"日蚀"
136	白花帝王球兰	034	短翅球兰	142	厚冠球兰"月影"
146	白花爪哇球兰	246	断叶球兰	072	厚花球兰
174	白玫瑰红球兰	216	钝叶球兰	102	护耳草
258	斑印球兰	070	多温球兰	260	花葶球兰
260	棒叶球兰		E	266	怀德球兰
028	贝拉球兰	084	恩格勒球兰	018	环冠球兰
026	本格尔顿球兰		F	104	黄花球兰
026	本格特球兰	248	反卷球兰	122	黄花休斯科尔球兰
096	鞭花球兰	092	斐赖迅球兰	268	黄结球兰
152	冰糖球兰	094	费氏球兰	104	黄色球兰
036	波特球兰	092	芬莱森球兰		J
032	布拉斯球兰	134	粉花帝王球兰	128	基心覆叶球兰
032	布拉轩球兰	122	粉花休斯科尔球兰	228	寄生球兰
	C	202	蜂出巢	246	截叶球兰
236	彩芯球兰	126	覆叶球兰	156	金边凹叶球兰
184	长叶球兰		G	160	金心凹叶球兰
172	橙花球兰	062	革叶球兰	050	景洪球兰
222	粗蔓球兰	112	格兰柯球兰	212	镜叶球兰
	D	116	格林球兰		K
070	达尔文球兰	064	冠花球兰	066	卡氏球兰
068	达奴姆球兰	064	冠状球兰	150	卡斯堡球兰
020	大花球兰	106	光叶球兰	148	堪雅酷玛瑞球兰
128	玳瑁叶球兰	126	龟壳叶球兰	042	康蓬那球兰
		058	桂叶球兰	162	柯氏球兰

L

166	兰氏球兰
114	烈味球兰
164	裂瓣球兰
202	流星球兰
254	柳叶球兰
186	卢卡球兰
220	卵叶球兰
180	罗比球兰
182	洛黑球兰
186	洛斯球兰
052	绿花球兰

M

196	玛丽球兰
190	麦季理斐球兰
190	麦氏球兰
244	毛萼球兰
194	美丽球兰
196	美林球兰
040	美叶球兰
076	密叶球兰
200	棉德岛球兰
170	棉毛球兰
038	缅甸球兰
200	民都洛球兰

N

024	南方球兰
204	瑙曼球兰
204	瑙珉球兰

Q

210	钱币球兰
210	钱叶球兰
226	琴叶球兰

048	青铜器球兰
060	丘生球兰
208	秋水仙球兰
046	球兰
108	球芯球兰
110	球序球兰

R

124	绒叶球兰

S

266	澈球兰
238	三脉球兰
086	珊瑚红球兰
030	双色球兰
252	苏格球兰
262	苏卡尔球兰
262	苏卡瓦球兰

T

218	甜香球兰
238	铁草鞋
082	椭圆叶球兰

W

272	瓦林球兰
270	瓦氏球兰
230	碗花球兰
276	威蔓尼球兰
274	威特球兰
276	薇蔓球兰
242	伪滨海球兰
048	尾状球兰
270	沃里克球兰

X

264	西藏球兰
256	希瑞球兰

256	锡亚球兰
056	纤毛球兰
178	线叶球兰
188	香花球兰
100	香水球兰
198	小花球兰
214	小棉球球兰
060	小丘球兰
110	小球球兰
254	谢普德球兰
154	心叶球兰
160	心叶球兰"内锦"
156	心叶球兰"外锦"
206	新喀里多尼亚球兰
122	休斯科尔球兰

Y

168	亚贝球兰
054	椰香球兰
070	蚁球球兰
080	异叶球兰
158	银斑凹叶球兰
158	银斑心叶球兰
148	印南球兰
250	硬叶球兰
058	玉桂球兰

Z

044	樟叶球兰
144	爪哇球兰
042	钟花球兰
066	孜然球兰

图书在版编目（CIP）数据

球兰：资源与栽培 / 黄尔峰, 王晖, 贾宏炎编著
. -- 沈阳：辽宁科学技术出版社, 2016.3
ISBN 978-7-5381-9698-6

Ⅰ. ①球… Ⅱ. ①黄… ②王… ③贾… Ⅲ. ①萝藦科—花卉—观赏园艺 Ⅳ. ① S68

中国版本图书馆 CIP 数据核字 (2016) 第 020424 号

出版发行：辽宁科学技术出版社
（地址：沈阳市和平区十一纬路 29 号　邮编：110003）
印　刷　者：利丰雅高印刷（深圳）有限公司
经　销　者：各地新华书店
幅面尺寸：215 mm×285 mm
印　　张：17.5
插　　页：4
字　　数：300 千字
印　　数：1～3000
出版时间：2016 年 3 月第 1 版
印刷时间：2016 年 3 月第 1 次印刷
责任编辑：杜丙旭
封面设计：许　佳
版式设计：吴　飞
责任校对：周　文

书号：ISBN 978-7-5381-9698-6
定价：328.00 元

联系电话：024-23284360
邮购热线：024-23284502
http://www.lnkj.com.cn